# Housing
# in
# Rural
# America

JOSEPH N. BELDEN & ROBERT J. WIENER *Editors*

# Housing in Rural America

## Building Affordable and Inclusive Communities

This publication is sponsored by

- California Coalition for Rural Housing Project
- Housing Assistance Council

**SAGE** Publications
*International Educational and Professional Publisher*
Thousand Oaks   London   New Delhi

*For information:*

SAGE Publications, Inc.
2455 Teller Road
Thousand Oaks, California 91320
E-mail: order@sagepub.com

SAGE Publications Ltd.
6 Bonhill Street
London EC2A 4PU
United Kingdom

SAGE Publications India Pvt. Ltd.
M-32 Market
Greater Kailash I
New Delhi 110 048 India

Printed in the United States of America

*Library of Congress Cataloging-in-Publication Data*

Main entry under title:

Housing in rural America: Building affordable and inclusive communities /
   edited by Joseph N. Belden, Robert J. Wiener.
         p. cm.
   Includes bibliographical references and index.
   ISBN 0-7619-1380-7 (acid-free paper)
   ISBN 0-7619-1381-5 (pbk.: acid-free paper)
   1. Housing, Rural—United States. I. Belden, Joseph N. II. Wiener, Robert J.
   HD7289.U6 H683 1998
   363.5'0973'091734—ddc21                                     98-40091

99   00   01   02   03   04   05   10   9   8   7   6   5   4   3   2   1

| | |
|---|---|
| *Acquiring Editor:* | Catherine Rossbach |
| *Editorial Assistant:* | Heidi Van Middlesworth |
| *Production Editor:* | Diana E. Axelsen |
| *Editorial Assistant:* | Patricia Zeman |
| *Typesetter/Designer:* | Marion Warren |
| *Cover Designer:* | Ravi Balasuriya |

# Contents

## Part IV
### Creative Solutions

# Foreword

*Gordon Cavanaugh*

For every person, a home—residence, housing, or shelter—is a defining and vital part of life. The better one's physical dwelling, the more likely it is that other areas of life will be successful. The focus of this book is the housing of Americans—particularly lower-income people—in small town and rural locales. Most of us think of rural areas in the United States in one or more of several ways: as peaceful and attractive places for a weekend visit or vacation, the land of quaintness and antiques; as places of abundant natural resources—farms, forests, and mines; or as pleasant retirement destinations, second homes intended as year-round residences in the future. Most of us today live in cities or their suburbs. The country is something out there, blending with and beyond the distant exurbs. We may, however, also think of rural and small town America as the place where we really would prefer to live, except for the unfortunate fact that we must live in cities or near them for employment.

This is the idyllic image of rural America. But there is another face to the countryside, the face of poverty. The historian Richard Hofstader wrote that "America was born in the country and moved to the city." Much of America did make this move, in migrations great and small since the late 19th century. But millions remained in rural places, and millions more in recent decades have returned to the countryside. For many, rural poverty has long been a reality and continues to be an affliction today. One of the most clearly evident manifestations of this poverty is rural housing.

Whether one finds rural housing to be inadequate, indecent, unfit, or unaffordable, the harsh reality is that the quest for shelter is a problem in the lives of millions of people in smaller communities. This important new book describes the rural millions who live in "cost-burdened" or substandard homes. The occupants may pay 50% or more of their incomes for housing, or they may live in shacks without running water. They may live in conditions that would be considered deplorable in a Third World country. They are in fact our country within a country.

When I became the first executive director of the Housing Assistance Council in 1971, it had only been four years since a Presidential commission called the rural poor "the people left behind." During my tenure as Administrator of the USDA, Farmers Home Administration, from 1977 to 1991, we tried to reach this population. Today, many are still behind. There are severe rural housing problems in every part of the nation, but conditions are especially bad in some high need and underserved areas: the Southwest border colonias, Appalachia, Indian country, the Mississippi Delta, the Ozarks, the Southeast, and farmworker regions. *Housing in Rural America* covers these conditions, analyzes public and private housing finance, and offers solutions to the rural housing crisis. This is a little known but important issue about which every American should be better informed. There are still many miles to go before we can say that we have provided, as the Housing Act of 1949 puts it, a decent home in a suitable living environment for every American family. The authors of this fine collection do much to advance their readers on that journey.

# Acknowledgments

This book would not have been written without the assistance of a number of people and institutions. The Fannie Mae Office of Housing Research-now part of the Fannie Mae Foundation-provided support and convened a 1995 research roundtable in which early versions of most of the chapters were presented. Our thanks to Jim Carr, Steve Hornburg, and Karen Danielsen of the Fannie Mae Foundation. The Ford Foundation for a number of years provided generous support to the Housing Assistance Council for a variety of tasks. This backing allowed some of HAC's staff to prepare chapters for the book. Thanks to Chris Page of Ford. Luz Rosas of HAC helped with production. Catherine Rossbach of Sage Publications had enough faith in the work to acquire it, and Diana Axelsen of Sage was an outstanding editor. Most important, the authors of all the chapters deserve our thanks for undertaking and completing the work through several drafts, updatings, and rewrites. And finally our families endured long nights and weekends of absenteeism. Our loving thanks to Lisa and Gabi, and to Tammy, Alicia, and Chris.

JOSEPH N. BELDEN
Housing Assistance Council

ROBERT J. WIENER
California Coalition for Rural
Housing Project

# Part I

# Conditions and Context

## Chapter 1

# The Context of Affordable
# Housing in Rural America

*Robert J. Wiener*
*Joseph N. Belden*

This book is about decent and affordable shelter in rural America, a little-known and often overlooked issue in housing policy. The rural poor and their housing conditions are not widely discussed or examined within the professional literature. This occurs in part because most housing policymakers, administrators, researchers, and advocates live in cities and take an urban-centric view—what some rural critics have called "metropolyanna" (Kravitz & Collings, 1986). Inner city problems are the main topics of most conversation about both affordable housing and poverty. But in fact, there is a broad spectrum of housing need, programs, new strategies, and practitioners in rural America. That is the subject of this book. This introductory chapter has two purposes: to define "rural" and to describe the state of rural housing and poverty in the United States. The balance of the book examines (in more detail) several especially distressed populations and regions, looks at housing credit needs and resources (both public and private) in rural areas, and concludes with several distinctly rural solutions to the housing problem.[1]

# The Meaning of "Rural"

"Rural" derives from the Latin word for "the country." In standard dictionaries, rural is anything having the characteristics of the country—rustic, pastoral, agrarian, or agricultural—as opposed to "urban" towns and cities. Typically, rural describes an area that is sparsely settled with wide open spaces, with its population engaged in resource-based occupations and its lifestyle and culture distinct from urban places. Wilson and Carr, in Chapter 12 of this book, refer to remoteness, low population density, and economic dependence on a single industry as defining characteristics.

Over time, demographic change, especially population growth, has challenged the conventional definitions. The more the country's population has grown, the more our understanding of rural has blurred. Whereas vast areas still are clearly rural, suburban sprawl in other areas has created a mosaic of regions that are alternately rural and urban in character. For example, rural pockets persist between new suburbs that have grown near urban centers. Within rural areas are pockets of "urbanity" in which small towns have grown into cities as urban "expatriates" move out beyond established suburbs. Conversely, some urban places have lost population and are more rural today than they were 20 to 30 years ago.

Commonly, "nonmetropolitan" or "nonmetro," as defined by the U.S. Bureau of the Census, is used in the professional literature as the statistical unit for analysis of rural conditions. "Rural" and "nonmetropolitan" often are used interchangeably but are not synonymous. The census defines rural areas as either open country or places of fewer than 2,500 residents. All other larger areas are urban. Nonmetropolitan areas are those counties that lie outside metropolitan statistical areas. Metro areas consist of counties with central cities of at least 50,000 residents and surrounding contiguous counties that are metropolitan in character. Metro areas can include rural places (fewer than 2,500 residents), and nonmetro areas include both rural and urban places (2,500 or more residents).

Some researchers, not wanting to be limited to the narrow census definition of rural or nonmetro, include the data on rural portions of metropolitan areas, also known as "rural metro" or "rural suburban." Others consider rural any housing located "outside urbanized areas," otherwise known as "nonurbanized areas." The census defines an "urbanized area" as an incorporated place and adjacent densely settled area with a population of at least 50,000 residents. Places outside urbanized areas are considered nonurbanized areas. Dolbeare, in Chapter 2 of this book, employs a unit of analysis that combines the census definitions of rural and another category called "other urban" (more than 2,500 but less than 50,000 residents).

The varying units of analysis in the literature are mirrored in the different definitions of rural in government programs of assistance. Rural housing programs use census population data and maps but expand the rural definition by setting higher population thresholds for program eligibility. Thus, populations in areas not defined as rural in the census (e.g., urban nonmetro) might qualify for assistance from rural programs. Unlike the census definition, which has remained constant since 1960, thresholds in these programs have tended to increase as rural areas have grown. However, there is no single definition embraced by all programs. This often has led to confusing and contradictory rules.

For purposes of this chapter, the main unit of analysis for rural is not the limited definition used by the census but rather nonmetropolitan areas. We also refer to the rural portions of metropolitan areas (rural metro) and compare data for rural nonmetro areas with data for rural metro areas where the comparisons highlight important nonmetro/metro differences. However, some of the rural data cited in this book might conform to different definitional frameworks.

## Pieces of the Problem: Conditions of Rural Housing and Poverty

The United States had 97.7 million occupied housing units in 1995, the latest year for which data are available. Table 1.1 shows that nonmetropolitan areas had 21.6 million occupied units, whereas rural metro areas had 13.2 million occupied units. Rates of homeownership were far higher in nonmetro and rural metro areas than in the central cities. Fully 74% of nonmetro and 82% of rural metro units were homeowner occupied, compared with 59% of urban and 49% of central city units. Generally, rural housing, including much rental housing, is predominantly in single-family units. In both nonmetro and rural suburban areas, approximately three of four occupied units were single-family detached homes. As Table 1.1 shows, almost 14% of nonmetro, 15% of rural suburban, and 18% of rural nonmetro residents lived in mobile homes. These proportions have been growing in recent years.

**TABLE 1.1** Occupied Housing Units in United States, 1995 (in thousands)

| | Total | Owner Occupied | Percentage Owner Occupied | Renter Occupied | One-Unit Detached | Mobile Homes |
|---|---|---|---|---|---|---|
| Total United States | 97,693 | 63,544 | 65.0 | 34,150 | 60,826 | 6,164 |
| Metropolitan areas | | | | | | |
| Central cities | 30,243 | 14,808 | 49.0 | 15,434 | 14,429 | 324 |
| Suburbs | 45,864 | 32,880 | 71.7 | 12,984 | 30,709 | 2,910 |
| Nonmetropolitan areas | 21,586 | 15,855 | 73.4 | 5,731 | 15,688 | 2,931 |
| Urban areas | | | | | | |
| Total | 70,683 | 41,696 | 59.0 | 28,987 | 40,431 | 1,741 |
| Nonmetropolitan | 7,914 | 4,908 | 62.0 | 3,005 | 5,343 | 430 |
| Rural areas | | | | | | |
| Total | 27,010 | 21,848 | 80.9 | 5,163 | 20,395 | 4,423 |
| Metro suburbs | 13,173 | 10,793 | 81.9 | 2,380 | 9,920 | 1,920 |
| Nonmetropolitan | 13,673 | 10,947 | 80.1 | 2,726 | 10,345 | 2,501 |

SOURCE: U.S. Bureau of the Census and U.S. Department of Housing and Urban Development (1997).

As Table 1.2 shows, many nonmetro and rural suburban households in 1995 lived in units that had moderate or severe problems in the physical conditions of the dwellings. However, on a proportional basis, this problem actually was greatest in central cities and rural nonmetro areas. Nonmetro areas, and particularly rural nonmetro places, were more likely to have higher proportions of units that lacked some or all plumbing, drew their water from wells, and had water leakages. Table 1.2 also shows that nonmetro units were smaller in size than suburban units, smaller even than central city units, and were less likely than suburban units to have two or more bathrooms.

Median incomes tended to be much lower in nonmetro areas in 1995, as Table 1.3 shows. Rural metro incomes were much higher than rural nonmetro incomes, a reflection of the growth of many rural suburbs as bedroom commuter communities. Approximately 23% of nonmetro households were "cost burdened." The housing costs of these occupied units consumed more than 30% of the incomes of their owners or renter households. Cost burden was higher in the cities.

**TABLE 1.2** Selected Unit Characteristics of Occupied Housing in United States, 1995

| | Total Units (in thousands) | Percentage With Severe or Moderate Physical Problems | Percentage Lacking Some or All Plumbing | Percentage With No Heating Equipment | Percentage With Dug Wells as Water Sources | Percentage With Leaking Roofs From Outside Structures | Median Square Feet of Unit | Percentage With Two or More Bathrooms |
|---|---|---|---|---|---|---|---|---|
| Total United States | 97,693 | 6.5 | 1.5 | 1.0 | 1.2 | 7.4 | 1,732 | 39.6 |
| Metropolitan areas | | | | | | | | |
| Central cities | 30,243 | 8.0 | 1.5 | 0.7 | <0.1 | 7.4 | 1,698 | 29.7 |
| Suburbs | 45,864 | 4.8 | 1.4 | 1.0 | 1.0 | 7.1 | 1,833 | 48.1 |
| Nonmetropolitan areas | 21,586 | 8.2 | 1.7 | 1.6 | 3.3 | 8.2 | 1,553 | 35.2 |
| Urban areas | | | | | | | | |
| Total | 70,683 | 6.5 | 1.4 | 0.8 | 0.1 | 7.1 | 1,762 | 37.0 |
| Nonmetropolitan | 7,914 | 8.5 | 1.1 | 0.9 | 0.1 | 7.2 | 1,551 | 28.2 |
| Rural areas | | | | | | | | |
| Total | 27,010 | 6.5 | 1.7 | 1.7 | 4.1 | 8.4 | 1,671 | 46.4 |
| Suburbs | 13,173 | 4.9 | 1.3 | 1.4 | 3.0 | 8.0 | 1,785 | 53.3 |
| Nonmetropolitan | 13,673 | 8.1 | 2.1 | 2.0 | 5.1 | 8.8 | 1,554 | 39.3 |

SOURCE: U.S. Bureau of the Census and U.S. Department of Housing and Urban Development (1997).

For a number of other indicators, there were clear differences between city and country in 1995. Among rural households, family size usually was larger and there generally was greater poverty as compared with urban areas. Nonmetro families also were more likely to include members who were older and had lower levels of education. And, as Table 1.3 also shows, nonmetro and rural households were less likely to be receiving food stamps and more likely to have some savings or investments as compared with households in the central cities. Rural households also were more likely to be close to, but above, the poverty threshold. Approximately 23% of rural households and 18% of urban households had incomes that placed them between 100% and 200% of poverty.

Table 1.4 shows that, despite need, few nonmetro households benefited from federal, state, or local housing assistance. Relative to the total nonmetro poverty population, the proportion of nonmetro families living in public housing was 14%, and the proportion of nonmetro families living in other federally subsidized units was 8%. By contrast, 23% of poor central city households were residents of public housing, and 14% lived in other federally subsidized units. The relative lack of subsidized units exacerbated the cost problem in nonmetro areas.

In nonmetro and rural places, financing for affordable housing often is difficult to find. As the data in Table 1.1 show, homeownership was the prevailing form of rural tenure in 1995. For rural Americans, however, mortgage credit generally comes at higher interest rates and shorter amortization periods than those required of borrowers in central cities and suburbs. Table 1.5 shows that in 1995, the median interest rates on primary mortgages (including government-subsidized mortgages) for owner-occupied homes were 8.7% in nonmetro areas, 8.3% in central cities, and 8.2% in suburban areas. The median terms of primary mortgages were 23 years in nonmetro areas, 30 years in central cities, and 29 years in suburban areas (U.S. Bureau of the Census & U.S. Department of Housing and Urban Development [HUD], 1997). It is not clear from the published data whether these terms (particularly the high interest rates) were the result of old loans that had not been refinanced. Scarcity of credit in more rural locations limits the ability of borrowers to refinance on more favorable terms.

These housing conditions are part of a larger social picture. One of the great secrets in debates on poverty is that the majority of the poor in the United States do not live in central cities. In 1995, there were 36.5 million persons in the United States living below the poverty level. Of the total, 16.3

**TABLE 1.3** Selected Household and Income Characteristics of Occupied Housing Units in United States, 1995

| | Total Units (in thousands) | Percentage Paying 30% or More of Income for Housing | Household Median Income (dollars) | Percentage With Less Than $25,000 Income Receiving Food Stamps | Percentage With No Savings or Investments | Median Age of Householder (years) | Percentage Householders With Less Than 9 Years of Education |
|---|---|---|---|---|---|---|---|
| Total United States | 97,693 | 30.4 | 31,416 | 7.1 | 23.1 | 46 | 7.8 |
| Metropolitan areas | | | | | | | |
|   Central cities | 30,243 | 36.4 | 27,859 | 10.3 | 30.3 | 44 | 8.4 |
|   Suburbs | 45,864 | 29.7 | 37,644 | 4.4 | 17.2 | 46 | 5.8 |
| Nonmetropolitan areas | 21,586 | 23.3 | 26,655 | 8.2 | 25.8 | 49 | 11.1 |
| Urban areas | | | | | | | |
|   Total | 70,683 | 33.0 | 31,157 | 7.6 | 24.2 | 45 | 7.2 |
|   Nonmetropolitan | 7,914 | 27.4 | 24,428 | 10.1 | 29.7 | 49 | 9.5 |
| Rural areas | | | | | | | |
|   Total | 27,010 | 22.7 | 32,027 | 5.8 | 20.4 | 48 | 9.3 |
|   Suburbs | 13,173 | 24.4 | 37,106 | 4.4 | 17.2 | 46 | 6.7 |
|   Nonmetropolitan | 13,673 | 21.0 | 27,904 | 7.1 | 23.6 | 50 | 12.0 |

SOURCE: U.S. Bureau of the Census and U.S. Department of Housing and Urban Development (1997).

**TABLE 1.4** Rent Subsidies in Occupied Housing Units in United States, 1995 (units and households in thousands)

| | Households Below Poverty Level | Units Owned by Public Housing Authorities | Percentage | Units With Other Federal Housing Subsidies | Percentage | Units With Other State or Local Housing Subsidies | Percentage |
|---|---|---|---|---|---|---|---|
| Total United States | 14,696 | 2,434 | 16.6 | 1,653 | 11.2 | 576 | 3.9 |
| Metropolitan areas | | | | | | | |
| Central cities | 5,925 | 1,380 | 23.3 | 810 | 13.7 | 293 | 4.9 |
| Suburbs | 5,004 | 521 | 10.4 | 556 | 11.1 | 179 | 3.6 |
| Nonmetropolitan areas | 3,767 | 533 | 14.1 | 287 | 7.6 | 104 | 2.8 |

SOURCE: U.S. Bureau of the Census and U.S. Department of Housing and Urban Development (1997).

**TABLE 1.5** Selected Characteristics of Primary Mortgages, Owner-Occupied Units, in United States, 1995

|  | Central Cities | Suburbs | Nonmetropolitan Areas |
|---|---|---|---|
| Median term of primary mortgage (years) | 30 | 29 | 23 |
| Current interest rate (percentage) | 8.3 | 8.2 | 8.7 |

SOURCE: U.S. Bureau of the Census and U.S. Department of Housing and Urban Development (1997).

million lived in central cities, 12.1 million lived in suburbs, and 8.1 million lived in nonmetropolitan areas (U.S. Bureau of the Census, 1996). Poverty in nonmetro America was, until recently, higher than poverty in the inner cities. Today, city poverty is higher, but for minorities and some subgroups, rural conditions actually are much worse. The popular assumption is that minorities make up most of America's poor. Actually, about two thirds of low-income people are white, at least according to census counts. A little-known fact is that it actually is white inner city poverty that raises the overall inner city poverty rate above that of nonmetro America. In nonmetro areas in particular, many low-income residents are white. However, the relatively smaller numbers of minorities in nonmetro and rural areas suffer very high poverty rates, much higher than those of whites. For African Americans and Hispanics, poverty actually is higher in nonmetro areas than in the central cities or suburbs. Female-headed families, the elderly, and children also have very high poverty rates in rural areas. Every state and region of the nation has rural poverty. This is not to suggest that the plight of the inner city poor is any less important than are rural concerns. But in popular and policy discussions, the media, and academia, the rural poor are mostly ignored.

## The Future

Clearly, there have been improvements in the housing situations of low-income rural and nonmetropolitan Americans in recent years. However, much remains to be done. The improvements have, in fact, roughly paralleled increases in spending for rural housing and other programs. For example, census and HUD data show that the number of nonmetro and rural suburban occupied units without complete plumbing facilities declined from about 2.2 million in 1970 to about 500,000 in 1995. Although it is possible that units lacking plumbing are undercounted, there clearly has been much progress. Over roughly the same period (i.e., primarily since the late 1960s), the U.S. Department of Agriculture housing programs have financed about 2 million units for low-income rural people. These programs, and those of HUD, deserve much of the credit for any improvements. Even with such progress, however, there are many families still living in shacks with dirt floors and outhouses, holes in the walls and ceilings, inadequate heating and water supplies, and high cost burdens. There are farmworkers who live in caves or boxes because there is no shelter. There are Indian reservations on which four of five homes are substandard. There are many nonmetropolitan counties in which 30% to 40% or more of the population lives below the poverty threshold. Rapidly increasing proportions of the nonmetro and rural populations live in mobile homes, while an older housing stock continues to decay. To alleviate these continuing problems, there must be an ongoing role for government as well as private and nonprofit initiatives. Innovative efforts such as creative public/private financing schemes, federal block grants (e.g., HUD's HOME and Community Development Block Grant programs), and the self-help housing program confound the popular belief that government is doomed to fail. Private approaches such as the expanding efforts of secondary market institutions, the rapid growth of nonprofit housing corporations, and the successes of community land trusts show that this diverse sector can and must play a vital role. The public and private victories may have been relatively small in some instances, but they are successes. The challenge is to ensure that these and other efforts flourish and multiply.

## Note

1. A number of the chapters in the book were presented as papers at a Fannie Mae Research Roundtable, Washington, D.C., October 1995. The roundtable was titled "A Home in the Country: The Housing Challenges Facing Rural America."

# Chapter 2

# Conditions and Trends
# in Rural Housing

*Cushing N. Dolbeare*

This chapter focuses primarily on problems of housing quality and affordability and on characteristics of households with those problems, using data from the biennial American Housing Surveys (AHS) conducted by the U.S. Bureau of the Census and U.S. Department of Housing and Urban Development (HUD). Most data used are from the 1995 AHS (U.S. Bureau of the Census & HUD, 1997), the most recent available, with some appropriate comparisons drawn from earlier years. In contrast to the decennial census, the AHS provides a wealth of data on housing characteristics, including housing quality. However, as a national survey, the sample size does not permit analysis of local data and conditions.

In comparison to either national housing data or urban housing conditions, relatively little analysis has been done of rural housing needs and how they differ from those in urban areas. The AHS provides information on seven different categories of residential location: central cities, urbanized suburbs, other urban suburbs, rural suburbs, urbanized nonmetro, other urban nonmetro, and rural nonmetro.

AUTHOR'S NOTE: A longer version of this chapter was prepared for the Housing Assistance Council under U.S. Housing and Urban Development Cooperative Agreement No. H-5971 CA.

Rural housing, as used in this chapter, is that found in America's small towns and open country. The closest census categories, and those used in this analysis, comprise housing that is located outside of central cities and other urbanized areas. In other words, rural housing includes both "rural" and "other urban" housing, whether or not it is located within or outside metropolitan areas. It excludes both central cities and urbanized areas, again whether or not they are within or outside metropolitan areas. (This differs from the usual definitions of rural and nonmetro; see Belden and Wiener's chapter in this book [Chapter 1].) According to the 1980 census definition, which still was used in the 1995 AHS, an urbanized area "comprised an incorporated place and adjacent densely settled (1.6 or more people per acre) surrounding area that together have a minimum population of 50,000" (U.S. Bureau of the Census & HUD, 1997, p. A2). "Other urban areas" consisted of places with 2,500 or more residents outside of urbanized areas. Rural housing is defined in the AHS as all housing "not classified as urban."

## General Characteristics of Rural Housing

In 1995, there were a total of 97.7 million occupied housing units in the United States. Of these, 60.5 million were urban, including 30.2 million in central cities, and 37.2 million (or 38%) were rural as we are using it in this study. Close to half of these rural units (44%) were located in metropolitan areas, with the remaining 56% outside of metropolitan areas. Approximately 70% were classified as rural by the census definition, with the remaining 30% classified as "other urban."

The reason why nearly half of all rural units, as we are using the term, were within metropolitan areas is in large part a function of the official definition of a metropolitan area—a city of at least 50,000 residents or an urbanized area of at least 50,000 residents with a total metropolitan population of 100,000 (75,000 in New England). Except in New England, all metropolitan areas are defined in terms of entire counties. Moreover, in addition to the county containing the main city, other entire suburban counties are included in metropolitan areas "if they are economically and socially integrated with the central city" (U.S. Bureau of the Census & HUD, 1997, p. A3). For example, much of the state of California, including most of the major agricultural areas, is metropolitan by the census definition.

# Urban/Rural Differences

Rural householders were far more likely than urban householders to be homeowners. Fully three quarters (76%) of rural householders were homeowners, compared with 58% of urban householders. Generally, rural housing, including much rental housing, was predominantly in single-family units; fully 73% was in single-family detached units, and another 3% was in single-family attached units. Only 11% of rural householders lived in structures with two or more units, which was less than one third the proportion of urban householders in multifamily housing. About one of eight rural householders lived in mobile homes, compared to only 2% of urban householders. Rural householders also were older, larger, more likely to be married, and generally poorer than were urban householders. More than half (56%) of rural householders were age 45 years or over, compared with 49% of urban householders. Only 21% of rural householders were under 35 years of age, compared with 28% of urban householders. Just over one fifth (22%) of rural households consisted of single persons living alone, compared with 26% of urban households. Fully 60% of rural householders were married, compared with only 49% of urban householders. Poverty rates in both rural and urban areas were similar, at about 15% of all households. However, rural households were more likely than urban households to be "near poor," with 22% of rural households and 18% of urban households having incomes between 100% and 200% of poverty.

Rural housing was less costly and bigger than urban housing. Rural households were about twice as likely as urban households to have monthly housing costs of less than $250 (25% of rural households vs. only 12% of urban households). At the other end of the scale, urban households were much more likely to pay more than $750 monthly for their housing (22% of rural households vs. 34% of urban households). Not surprisingly, therefore, rural households were somewhat less likely than urban households to have high housing cost burdens. Even so, 22% of rural households and 31% of urban households paid more than 30% of their incomes for housing. Only 5% of rural housing units had fewer than four rooms, compared with 13% of urban units. Somewhat surprisingly, rural housing appeared to have no greater problems of housing quality than did urban housing. The traditional indicator of housing quality, lack of complete plumbing facilities, shows that less than 2% of either rural or urban units lacked this essential. There also was little difference in the AHS composite indicator of quality—physical adequacy of the unit. Fully 93% of rural units and 94% of urban units were classified as adequate.

Although rural America contains many areas with substantial numbers of Hispanics, African Americans, and/or American Indians, overall, rural residents were more likely than urban residents to be white. Based on AHS data, fully 88% of rural householders were white, compared with 70% of urban householders. African Americans comprised 15% of urban households but only 6% of rural households. Hispanics accounted for 10% of urban households but only 4% of rural households. Finally, 2% of rural households belonged to other minorities, primarily American Indians, and 4% of urban households belonged to other minorities, primarily Asians or Pacific Islanders. These disparities probably can be accounted for by two factors: the 20th-century migration of African Americans from the rural South to large cities and the tendency of immigrants to settle in urban areas.

Based on AHS data, more than one quarter (28%) of rural households had major housing problems. Of the 10.4 million households with problems, 4.9 million had moderate cost burdens, paying from 30% to 50% of income for housing; 3.1 million had severe housing cost burdens, paying more than half their incomes; 2.6 million lived in housing classified as moderately inadequate[1] or seriously inadequate;[2] and 0.7 million were overcrowded with more than one person per room. There is some duplication in these counts, as 0.9 million households had more than one of these problems, generally both high costs and inadequate quality.

## Trends

Using data from the AHS from 1985 through 1995, it is possible to identify recent trends and changes in the housing stock. Overall, the number of occupied housing units increased by 13% from 1985 to 1995, from 86.3 million to 97.7 million. Most of this increase occurred between 1985 and 1991. Rural growth was much higher than that of central cities and nearly as high as that in the suburbs. Rural units increased from 32.0 million in 1985 to 37.2 million in 1995, a growth of 16%. Central cities grew by 5%, while suburbs grew by 19%.

Most rural growth was in rural parts of metropolitan areas. In the period from 1985 to 1995, the number of metropolitan area households living outside of "urbanized" or "other urban" areas grew by 28%, from 9.9 million to 12.6 million households. Comparable nonmetro areas grew by 17%. Other urban metro areas expanded by 14%, but other urban nonmetro areas increased by only 2%. The net result is that rural households grew from 37% of all households in 1985 to 38% in 1995.

## Housing Affordability

The gap between what people can afford to pay and the cost of housing, including utilities and (for homeowners) taxes and maintenance, is the major housing problem facing this country. In 1995, 11.1 million households had severe cost burdens, paying more than 50% of their incomes for housing. Of these households, 3.1 million lived in rural areas. Another 15.4 million households had moderate cost burdens, paying more than 30% but not more than 50% of their incomes for housing costs. Of these households, 4.9 million lived in rural areas. In all, 8.0 million households, or 22% of all rural households, suffered from high housing cost burdens. Shocking as this figure may be, it is lower than cost burdens in urban areas. One third of central city residents (33%) paid more than 30% of their incomes for housing costs, as did 29% of suburban residents. See Figure 2.1 for incidence of rural cost burdens by tenure.

Although incomes in rural areas tend to be somewhat lower than those in urban areas, housing costs also are lower. Therefore, rural households are less likely to face affordability problems than are households in central cities or urbanized areas. In 1995, the median family income in rural areas was $29,200 annually, compared with $31,800 in urban areas as a whole. However, the median income in central cities, $26,000, was far lower than the $38,000 median income in suburbs. However, median monthly housing costs in rural areas were only $408, compared with $589 in urban areas. Central city monthly housing costs were $521, compared with suburban costs of $672.

Housing cost burdens generally are measured as a percentage of income on what has become a slowly sliding scale. In the early days of the public housing program, housing costs higher than 20% of income were considered burdensome. In the late 1960s and early 1970s, 25% of income became the dividing line. In the early 1980s, the cost burden threshold was raised again—to 30% of income. Since then, HUD has defined moderate cost

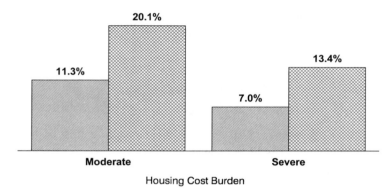

**Figure 2.1.** Housing Cost Burdens by Tenure in Rural Areas, 1995

burdens as those between 30% and 50% of income and severe cost burdens as those higher than 50% of income.

Percentage of income paid for housing is, at best, a rough measure of affordability. Its use has become widespread for several reasons. First, it is relatively simple to grasp and calculate. Second, 30% of income has become the norm for households living in subsidized housing. Clearly, however, the proportion of income that is affordable for housing depends both on one's income level and on other basic needs. Larger households, with more mouths to feed, can afford less than can smaller households at the same income level. High-income people could, at least in theory, afford to spend half their incomes for housing and still pay for other necessities.

Arguably, the lower a household's income, the smaller the proportion it can afford for housing. In practice, however, exactly the opposite happens— the lower the income, the higher the proportion that goes for housing costs. In 1995, fully 40% of rural households with incomes below $10,000 paid more than half of these meager incomes for housing, and another 25% had moderate cost burdens. Only 35% had costs of 30% of income or less. Indeed, 85% of rural householders who reported paying more than 50% of their incomes for housing costs had incomes below $20,000. At the other end of the scale, only 4% of rural households with incomes above $60,000 had moderate cost burdens, and less than 0.5% reported severe cost burdens.

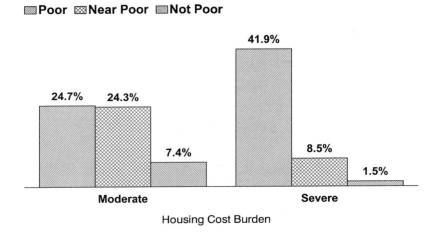

**Figure 2.2.** Rural Households With Housing Cost Burdens by Poverty Status, 1995

Because it takes household size into account, the poverty line probably is a better measure of ability to pay for housing. In 1995, fully 42% of poor rural households had severe cost burdens, and another 25% had moderate cost burdens. Only 30% had affordable housing by the "30% of income" housing cost standard. Cost burdens dropped substantially for near-poor households, that is, those with incomes between 100% and 200% of the poverty level. In this category, 9% of the households had severe cost burdens, and another 24% had moderate cost burdens. Severe cost burdens were almost negligible for households with incomes above 200% of the poverty level; only 2% had severe cost burdens, although 7% had moderate cost burdens (see Figure 2.2).

Rural renters were far more likely than homeowners to have severe cost burdens. Approximately 13% of all rural renters had severe cost burdens, and another 20% had moderate cost burdens. However, only 7% of rural home-owners had severe burdens, and 11% had moderate burdens. Among home-owners, 9% of those with subsidized mortgages reported having severe cost burdens, and another 15% had moderate cost burdens. Yet, because 77% of these subsidized homeowners did not have high cost burdens, it is clear that the subsidy assistance has made a major difference to them.

High cost burdens were primarily a factor of low income rather than of high housing costs. Just over half (51%) paid more than $500 monthly for their housing. Approximately 13% of households with severe cost burdens paid less than $250 monthly for their housing costs, and another 36% paid $250 to $500. In addition, 52% of households with moderate cost burdens paid more than $500 monthly, 16% paid less than $250, and another 32% paid $250 to $500.

## Substandard Housing

The AHS provides a wealth of data on physical housing characteristics and also compiles significant data to measure moderate or severe housing inadequacy based on the following five problem areas: plumbing, heating, upkeep, hallways, and electric. Because this indicator is the most widely used general measure of housing quality, we focus on it first.

By these measures, 6.4 million households were living in inadequate units in 1995. Residents of central cities and rural areas were far more likely than suburban residents to be in inadequate units. Approximately 7% of rural households and 8% of central city households were in inadequate housing, compared with only 4% of suburban households.

Significantly, more than a quarter (27%) of rural households living in poor housing also paid more than 30% of their incomes for their units. Renters were far more likely than homeowners to live in inadequate housing. Approximately 10% of rural renters, 6% of suburban renters, and 10% of central city renters were in inadequate housing, compared with 6% of rural homeowners, 5% of central city homeowners, and 4% of suburban homeowners. Yet, because homeowners predominate in rural areas, 1.7 million of the 2.6 million rural households in inadequate housing were homeowners, whereas 1.6 million of the 2.4 million central city households in inadequate housing were renters.

Housing inadequacy is not quite as closely related to poverty as are housing cost burdens. Approximately 33% of rural households in severely inadequate housing had incomes below the poverty level, and 26% were near poor. By comparison, 32% of moderately inadequately housed rural households were poor, and 26% were near poor.

Black rural households were more than twice as likely as white households to live in seriously inadequate housing and were more than four times as likely to live in moderately inadequate housing. Approximately 17% of black households, but only 4% of white households, lived in moderately inadequate housing, whereas 5% of black households and 2% of white households were in severely inadequate housing. Among Hispanics, 3% lived in severely inadequate housing, and 10% were in moderately inadequate housing.

It is clear from the definition of inadequate housing that this general concept is a fairly loose proxy for substandard housing. For example, 0.7 million of the 1.2 million rural units with broken plaster or peeling paint were classified as "adequate," as were 93% of the rural units reporting basement leaks, 53% of the units reporting open cracks or holes in walls or floors, 82% of those reporting inadequate heat, 85% of those reporting leaks from inside the building, 89% of those reporting leaks from outside the building, and 62% of those reporting rats.

## Overcrowded Housing

Overcrowding is far less of a problem than is either high cost or quality. Overall in 1995, approximately 2.6 million households, or 2.6% of all households, had more than one person per room. The incidence of overcrowding was a shade below 2% in rural areas and just above 2% in suburban areas, whereas it was 4% in central cities.

More than half of the 696,000 overcrowded rural households were homeowners. Fully 99% had at least one child, compared with 38% of uncrowded households. Approximately 67% of the householders were married. Fully 93% of overcrowded rural households had five or more persons in their household, and 52% of their housing units had at least five rooms. In addition, 41% of overcrowded rural households had incomes below the poverty line, and another 37% were near poor. Only 23% had incomes above 200% of the poverty level. By contrast, 14% of uncrowded households were poor, 23% were near poor, and 64% had incomes above 200% of the poverty level. In addition, 11% of crowded rural households had severe housing cost burdens, and another 14% had moderate cost burdens. Finally, 10% lived in severely or moderately inadequate housing.

## Housing Problems of African American and Latino Households

The AHS provides data on a number of racial and ethnic groups—African Americans, Hispanics, American Indians, Asians/Pacific Islanders, and "other." However, the sample sizes for the last three groups are so small that there is a wide margin of error in the estimates. Still, although only 12% of rural households belonged to minority groups in 1995, they accounted for a significant proportion of all minorities. One fifth of all Latino and African American households lived in rural areas, as did 57% of Native Americans and 9% of Asians/Pacific Islanders. The 2.4 million African American households represented the largest minority group in rural areas, comprising 6% of all rural households. The 1.5 million Latino households comprised 4.2% of all rural households. American Indians, Asians/Pacific Islanders, and others together constituted 1.5% of all rural households.

The housing characteristics of minority households in rural areas reflect a combination of some of the significant minority characteristics, such as low income and low rates of homeownership, and rural characteristics. Significantly, homeownership rates for minority households in rural areas were significantly higher than they were elsewhere. The homeownership rate for all African American households was 44%, but in rural areas it was 61%. Similarly, the overall Hispanic homeownership rate was 42%, but it was 58% in rural areas. The homeownership rate for American Indians was 48% overall but was 54% in rural areas. For Asian/Pacific Islander households, it was 53% overall and 73% in rural areas. Rural blacks and Hispanics also were far more likely than their urban counterparts to be married. Overall, only 32% of blacks were married, whereas 39% of rural blacks were married. Because a far higher proportion of Latinos marry, the differential is less dramatic; that is, 60% of rural Latinos were married, compared with 54% of Latinos who were married overall.

Rural Hispanics and blacks also were less likely than those living in urban areas to have housing affordability problems. Fully 68% of blacks in rural areas paid 30% or less of their incomes for housing costs, and only 13% had severe housing cost burdens, compared with 22% of urban blacks who had severe cost burdens. Similarly, 73% of rural Hispanics did not have affordability problems, compared with 57% of urban Hispanics. Only 12% of rural Hispanics had severe cost burdens, compared with 21% of urban Hispanics. White households were far less likely than minority households to have housing affordability problems. Only 8% of white households had severe cost burdens, and another 13% had moderate cost burdens. By contrast, 13% of African American households and 12% of Hispanic households had severe cost burdens. Moderate cost burdens were a problem for 19% of African American households and 17% of Hispanic households.

Rural blacks, however, were far more likely to live in inadequate units. Nearly one quarter of rural black households (22%) lived in moderately or severely inadequate units, compared with 12% of urban black households. Overall, the condition of housing occupied by Latinos was better than that occupied by blacks but still was far worse than that occupied by whites. The 13% of inadequately housed rural Hispanic households was only slightly higher than the 11% of inadequately housed urban Hispanics. Crowding was a major problem faced by Hispanics. Probably because rural units are larger, the rate for rural Hispanic households was 13%, slightly below the urban rate of 15%.

## Housing Problems of Rural Elderly and Single-Parent Households

### Elderly

A total of 8.9 million rural householders were age 65 years or over in 1995. This was more than a quarter (27%) of all rural householders. Fully 85% of these elderly householders were homeowners. More than half were poor or near poor, with 21% below the poverty line and 31% between 100% and 200% of the poverty line. Overcrowding was not a problem for elderly households, as 91% consisted of only one or two persons and 93% of these householders lived in units with four or more rooms. Fully 80% were in single-family structures. In addition, 28% had problems of affordability, quality, or both. Furthermore, 8% had severe cost problems, and 13% had moderate cost problems. Finally, 7% were in inadequate units.

Fully 96% had social security income, and 5% received supplemental security income (SSI). More than a quarter (26%) had income from wages or other earnings. In addition, 29% of elderly renters reported living in subsidized housing. The majority (53%) paid less than $250 monthly for housing costs.

### Single-Parent Households

One tenth of all rural householders (approximately 3.9 million) were single parents, and 43% of these single parents had at least one child under

6 years of age. Just under half (48%) were renters. The majority (53%) lived in housing built since 1978, when lead-based paint was banned, but 28% lived in units built before 1950, which are likely to contain lead hazards dangerous to children's health. Nearly half (47%) of these households had one or more housing problems. In addition, 21% had moderate cost burdens, and 17% had severe cost burdens. Furthermore, 10% had housing quality problems, and nearly half either were crowded or had housing cost burdens. Finally, 6% were overcrowded, and 35% of these householders rated their units as either fair or poor.

Among single-parent households, 34% had incomes below the poverty level, and another 29% were near poor. In addition, 21% reported income from welfare or SSI, whereas 17% had social security income. By contrast, 81% reported at least some income from wages or earnings, and 42% received food stamps. Finally, 27% of renters lived in subsidized housing, and 15% of homeowners reported having subsidized mortgages.

## Subsidized Housing
## Availability Compared With Need

Federal housing assistance is an entitlement only for homeowners who can deduct their property taxes and mortgage interest from taxable income. The Office of Management and Budget has estimated the revenue loss to the federal Treasury from these entitlements in 1995 at $63 billion. The Congressional Joint Committee on Taxation estimated that 89% of the 1995 benefits of the mortgage interest and property tax deduction went to homeowners with incomes above $50,000. Because rural homeowners are poorer than urban homeowners, the result of this is that rural areas, with 44% of all homeowners, received only 32% of the benefits of these tax deductions in 1995.

In addition to these benefits, capital gains exclusions and deferrals added another $19 billion in benefits to homeowners, and investor benefits resulted in a revenue loss of $11 billion. Thus, total housing-related tax expenditures cost the Treasury $94 billion. No breakdown is available of the income distribution of these tax benefits, but it is highly unlikely that the proportion is as high as the proportion of mortgage interest and property tax deductions benefiting rural residents.

However, if one assumes, for the sake of analysis, that rural areas receive 32% of all housing-related tax benefits, then the total amount in 1995 would have been $30 billion, well above the $28 billion total of all fiscal year 1995 low-income housing outlays for HUD, the Farmers Home Administration (FmHA, now the Rural Housing Service [RHS]), and other federal housing programs. This amount probably would be sufficient to fund a substantial increase in the supply of affordable housing.

The 1995 AHS found that a total of 1.3 million rural renter households were living in federally subsidized housing (under HUD and FmHA programs). This was 25% of all subsidized renter households, exactly the same proportion as rural renters bear to the total number of renter households. Approximately 15% of all rural renters lived in federally assisted housing, as did 18% of central city renters and 12% of suburban renters.

HUD currently has no programs comparable to the RHS Section 502 low-interest homeownership program. Approximately 2% of rural homeowners with mortgages sampled in 1995 reported that their primary mortgage insurance was with FmHA, and 9% of rural homeowners with mortgages reported that they had obtained low-cost mortgages through the federal government. However, this latter figure should be viewed with skepticism given that 10% of central city homeowners also said that they had received low-cost mortgages.

The contrast between needs and current levels of housing assistance is stark. If an appropriate approximate measure of the number of rural households still needing housing is the number of poor and near-poor households with housing problems, then 3.0 million rural renters and 4.2 million homeowners have significant housing problems that they currently are unable to solve. If the federal government does not assist them in this era of shrinking state and local commitment to dealing with problems of low-income people, then there is little likelihood that they will ever get help.

## Conclusion

Rural housing problems are both substantial and solvable. As the forgoing analysis shows, they are comparable in urgency to those of central cities, although they have received far less attention. This is largely because although their scale is great, their density is low, so they do not appear as overwhelming as the housing and related poverty and lack of opportunity in urban areas.

Greater attention to rural housing is urgently needed. Without it, millions of families will be hurt, perhaps irreparably. Addressing rural housing needs calls for a change of national priorities and a recognition of the fundamental importance of decent homes and suitable living environments for all families in America.

# Notes

1. A housing unit is classified as "moderately inadequate" (with "moderate/severe physical problems" as defined by the AHS) if it has any of the following five problems but none of the severe problems (see Note 2):

   *Plumbing:* having the toilets all break down at once, at least three times in the last 3 months, for at least 6 hours each time;

   *Heating:* having unvented gas, oil, or kerosene heaters as the main source of heat (these give off unsafe fumes);

   *Upkeep:* having any three of the six upkeep problems mentioned under severe (see Note 2);

   *Hallways:* having any three of the four hallway problems mentioned under severe (see Note 2); or

   *Kitchen:* lacking a sink, range, or refrigerator, all for the exclusive use of the unit. (U.S. Bureau of the Census & HUD, 1997, p. A14)

2. A housing unit is classified as "seriously inadequate" (with "moderate/severe physical problems" as defined by the AHS) if it has any of the following five problems:

   *Plumbing:* lacking hot piped water or a flush toilet, or lacking both bathtub and shower, all for the exclusive use of the unit;

   *Heating:* having been uncomfortably cold the previous winter, for 24 hours or more, because the heating equipment broke down, and it broke down at least three times the previous winter, for at least 6 hours each time;

   *Upkeep:* having any five of the following six problems: leaks from outdoors, leaks from indoors, holes in the floor, holes or open cracks in the walls or ceilings, more than 1 square foot of peeling paint or plaster, or rats in the previous 90 days;

   *Hallways:* having all of the following four problems in public areas: no working light fixtures, loose or missing steps, loose or missing railings, and no elevator; or

   *Electric:* having no electricity or all of the following three electric problems: exposed wiring, a room with no working wall outlet, or three blown fuses or tripped circuit breakers in the previous 90 days. (U.S. Bureau of the Census & HUD, 1997, pp. A13-A14)

# Part II

People and Places

## Chapter 3

# Affordable Housing in the Rural South

*Jacquelyn W. McCray*

Rural housing in the South suggests visions of run-down shacks on dirt roads or of grand plantation homes with immaculate gardens. Both scenarios still are true but have diminished in quantity and visibility. Substantial improvements in rural housing quality in the region have been made in recent decades. Amenities once reserved for the wealthy, such as bathrooms and air conditioning, are now taken for granted.

Housing problems of the rural South captured the attention of scholars and political advocates in the 1940s and 1950s. In subsequent decades, much of the current knowledge base in southern rural housing, including the impact of various socioeconomic factors on housing, grew out of university and government accounts. An assessment of rural housing needs by Day (1978) reported that low-income, elderly, and nonwhite households occupy a large proportion of housing that is lacking complete plumbing, overcrowded, and in need of major repairs.

A few years later, Morris and Winter (1982) reported that the quality of rural housing, although a problem in all regions of the country, was consistently worse in the South than in the nation as a whole. Their study noted the location of a greater proportion of substandard units and units with severe structural defects in the rural South as compared with the North (urban or rural) or urban South. In the mid- to late 1980s, other researchers (Kravitz & Collings, 1986; Sindt & Guy, 1985) helped clarify regional and urban/rural differences in housing problems. Many acknowledged the need for adequate and affordable housing in most rural areas of the country but cited special problems in the South. A general consensus was that the extent of housing problems in rural areas often is less visible and more neglected.

Historically, inadequate plumbing placed rural southerners at greater risks of substandard housing. A mid-1980s analysis of American Housing Survey data found that three fourths of all households living in substandard housing resided in the South (Lazere, Leonard, & Kravitz, 1989). At that time, 42% of nonmetro housing was located in the South, but more than three fourths of rural households in substandard units were in the region. According to Lazere et al. (1989), poor rural southerners were most likely to live in housing with inadequate waste disposal (79%), inadequate water supplies (52%), inadequate heating equipment (61%), evidence of rats (93%), holes in the floor (73%), and cracks in the walls (53%) as well as without complete bathrooms (62%).

Rural housing concerns traditionally have focused on structural conditions or housing quality. In recent years, attention also has turned to affordability and availability issues. Population data indicate that many rural areas in the South grew economically and in population base between 1970 and 1980. New settlement patterns (reverse migration) raised questions concerning the adequacy of existing services including housing. In many rural communities, new demands for water and sewerage and for other economic and social services increased in the past two decades.

## The Current Situation

The 1990 census and the 1995 American Housing Survey provide an excellent opportunity to assess the status of rural housing in the South including its standing relative to other regions of the country. Table 3.1 provides a general comparison of household and housing characteristics in the South and nonmetro South with the nation as a whole in 1995. The nonmetropolitan South had 9.2% of all occupied housing units in the United States but also

**TABLE 3.1**  Characteristics of Housing Units and Households in United States and South, 1995 (in thousands)

| | United States, All Areas | South | Nonmetro South | Nonmetro South as Percentage of United States |
|---|---|---|---|---|
| All occupied units | 97,693 | 34,236 | 9,027 | 9.2 |
| Owner occupied | 63,544 | 22,959 | 6,752 | 10.6 |
| Percentage owner occupied | 65.0 | 67.1 | 74.8 | — |
| Renter occupied | 34,150 | 11,277 | 2,275 | 6.7 |
| Mobile homes | 6,164 | 3,216 | 1,623 | 26.3 |
| Physical problems | | | | |
| Severe | 2,022 | 648 | 225 | 11.1 |
| Moderate | 4,348 | 2,296 | 921 | 21.2 |
| Black householders | 11,773 | 6,295 | 1,444 | 12.3 |
| Hispanic householders | 7,757 | 2,686 | 299 | 3.9 |
| Elderly householders | 20,841 | 7,231 | 2,400 | 11.5 |
| Below poverty line | 14,695 | 5,725 | 1,822 | 12.4 |
| Received food stamps | 6,924 | 2,689 | 901 | 13.0 |
| Median household income (dollars) | 34,416 | 29,309 | 25,130 | — |

SOURCE: U.S. Bureau of the Census and U.S. Department of Housing and Urban Development (1997).

had 21.2% of the occupied homes with moderate physical problems, 11.1% of the units with severe physical problems, a remarkable 26.3% of the nation's mobile homes, 12.4% of the households living below the poverty line, and 13.0% of the households receiving food stamps.

The nonmetro South had 11.5% of the nation's elderly-occupied housing units and 12.3% of the African American households. Homeownership also was more prevalent in the nonmetro South than in any other region. In the 16 southern and border states, 30.2% of the population lived in rural areas.[1] The 1990 census shows that the poverty rate in the rural South was 22.2%. Some southern states had very high rural poverty and substantial rural populations, as shown here:

|                | Percentage of<br>Population That Is Rural | Percentage of Rural<br>Population With Incomes<br>Below the Poverty Line |
|----------------|-------------------------------------------|--------------------------------------------------------------------------|
| Alabama        | 40.2                                      | 18.3                                                                     |
| Arkansas       | 47.1                                      | 19.0                                                                     |
| Kentucky       | 49.0                                      | 22.0                                                                     |
| Louisiana      | 32.1                                      | 23.1                                                                     |
| Mississippi    | 53.0                                      | 26.0                                                                     |
| North Carolina | 50.7                                      | 12.4                                                                     |
| South Carolina | 46.0                                      | 15.5                                                                     |
| West Virginia  | 64.2                                      | 21.1                                                                     |

Tables 3.2 and 3.3 report on substandard units in the South. This is housing that lacks complete plumbing, is overcrowded, or is both lacking plumbing and overcrowded. Most researchers and housing advocates agree that this definition is wholly inadequate in defining the condition of the nation's housing stock. Shifts in housing quality measurements have occurred over the past several decades as census methodology and data collection strategies have changed. Enumerator evaluations of structural quality have not been reported since 1960. Although the 1990 census report provides the most comprehensive assessment of housing quality nationwide, the actual number of substandard units is substantially higher than the census figures. Data do not include measures of structural soundness, rodent or insect infestation, electrical resources, insulation, and other relevant variables.

**Plumbing Criterion**

Across the South, 410,498 units (1.36% of total) lacked complete kitchen facilities, and more than 241,000 of these units were located in rural areas (Table 3.2). Of the 345,000 units lacking complete plumbing, 242,509 were located in rural areas. Overall, 2.54% of rural units in the South lacked complete plumbing. This percentage was five times greater than the percentage of urban units with plumbing deficiencies. Rural units in Kentucky, Virginia, and West Virginia had higher rates of plumbing deficiencies (4.84%, 4.38%, and 3.55%, respectively) than did rural units in other states.

*(text continues on p. 37)*

**TABLE 3.2** Substandard Housing Units in South Region by Rural and Urban: Units Lacking Complete Kitchen and Plumbing Facilities, 1990

| Subregion and State | Housing Units Lacking Complete Kitchen Facilities | | | | | | Housing Units Lacking Complete Plumbing Facilities | | | | | |
| --- | --- | --- | --- | --- | --- | --- | --- | --- | --- | --- | --- | --- |
| | Total | | Rural | | Urban | | Total | | Rural | | Urban | |
| | No. | Percentage[a] | No. | Percentage[a] | No. | Percentage[a] | No. | Percentage[b] | No. | Percentage[b] | No. | Percentage[b] |
| South Atlantic[c] | 171,166 | 1.09 | 100,053 | 1.77 | 71,113 | 0.61 | 157,926 | 1.12 | 112,057 | 2.24 | 45,869 | 0.40 |
| Delaware | 1,933 | 0.67 | 832 | 0.85 | 1,101 | 0.57 | 1,160 | 0.47 | 744 | 1.12 | 416 | 0.23 |
| Maryland | 10,796 | 0.57 | 4,594 | 1.37 | 6,202 | 0.40 | 10,206 | 0.58 | 5,308 | 1.74 | 4,898 | 0.34 |
| Virginia | 33,097 | 1.33 | 24,458 | 3.18 | 8,639 | 0.50 | 35,788 | 1.56 | 30,003 | 4.38 | 5,785 | 0.36 |
| West Virginia | 18,276 | 2.34 | 15,412 | 3.18 | 2,864 | 0.97 | 15,972 | 2.32 | 14,925 | 3.55 | 1,047 | 0.39 |
| North Carolina | 33,774 | 1.20 | 24,058 | 1.70 | 9,716 | 0.69 | 33,192 | 1.32 | 27,743 | 2.26 | 5,449 | 0.42 |
| South Carolina | 16,121 | 1.13 | 10,805 | 1.73 | 5,316 | 0.66 | 16,626 | 1.32 | 12,715 | 2.31 | 3,911 | 0.55 |
| Georgia | 24,014 | 0.91 | 13,759 | 1.47 | 10,255 | 0.60 | 22,921 | 0.97 | 15,443 | 1.85 | 7,478 | 0.49 |
| Florida | 33,155 | 0.54 | 6,135 | 0.67 | 27,020 | 0.52 | 22,061 | 0.43 | 5,176 | 0.71 | 16,885 | 0.38 |
| East South Central | 96,357 | 1.60 | 72,162 | 2.71 | 24,195 | 0.70 | 95,907 | 1.76 | 81,020 | 3.39 | 14,887 | 0.48 |
| Kentucky | 31,161 | 2.07 | 25,852 | 3.64 | 5,309 | 0.67 | 33,623 | 2.44 | 30,921 | 4.84 | 2,702 | 0.36 |
| Tennessee | 25,708 | 1.27 | 17,984 | 2.33 | 7,724 | 0.62 | 23,840 | 1.29 | 19,438 | 2.78 | 4,402 | 0.38 |

*(continued)*

33

**TABLE 3.2** Continued

| Subregion and State | Housing Units Lacking Complete Kitchen Facilities | | | | | | Housing Units Lacking Complete Plumbing Facilities | | | | | |
|---|---|---|---|---|---|---|---|---|---|---|---|---|
| | Total | | Rural | | Urban | | Total | | Rural | | Urban | |
| | No. | Percentage[a] | No. | Percentage[a] | No. | Percentage[a] | No. | Percentage[b] | No. | Percentage[b] | No. | Percentage[b] |
| Alabama | 22,089 | 1.32 | 14,775 | 2.25 | 7,314 | 0.72 | 20,819 | 1.38 | 15,812 | 2.74 | 5,007 | 0.54 |
| Mississippi | 17,399 | 1.72 | 13,551 | 2.60 | 3,848 | 0.79 | 17,625 | 1.93 | 14,849 | 3.18 | 2,776 | 0.63 |
| West South Central | 142,975 | 1.38 | 69,145 | 2.32 | 73,830 | 0.92 | 91,337 | 0.99 | 49,432 | 1.98 | 41,905 | 0.51 |
| Arkansas | 16,376 | 1.64 | 11,842 | 2.56 | 4,534 | 0.84 | 13,030 | 1.46 | 10,328 | 2.56 | 2,702 | 0.55 |
| Louisiana | 22,710 | 1.32 | 10,733 | 2.01 | 11,977 | 1.01 | 14,318 | 0.95 | 8,335 | 1.83 | 5,983 | 0.57 |
| Oklahoma | 19,065 | 1.36 | 9,762 | 2.18 | 9,303 | 0.97 | 7,145 | 0.59 | 4,741 | 1.28 | 2,404 | 0.29 |
| Texas | 84,824 | 1.21 | 36,808 | 2.51 | 48,016 | 0.87 | 56,844 | 0.94 | 26,028 | 2.23 | 30,816 | 0.63 |
| Total South region | 410,498 | 1.36 | 241,360 | 2.27 | 169,138 | 0.74 | 345,170 | 1.29 | 242,509 | 2.54 | 102,661 | 0.46 |

SOURCE: Housing Assistance Council (1994a).
a. Percentage of all housing units.
b. Percentage of all occupied housing units.
c. District of Columbia excluded (data not available).

**TABLE 3.3** Substandard Housing Units in South Region by Rural and Urban: Overcrowded and Substandard Units, 1990

| Subregion and State | Overcrowded Housing Units | | | | | | Total Substandard Housing Units | | | | | |
| --- | --- | --- | --- | --- | --- | --- | --- | --- | --- | --- | --- | --- |
| | Total | | Rural | | Urban | | Total | | Rural | | Urban | |
| | No. | Percentage[a] | No. | Percentage[a] | No. | Percentage[a] | No. | Percentage[a] | No. | Percentage[a] | No. | Percentage[a] |
| South Atlantic[b] | 562,413 | 3.14 | 152,139 | 3.01 | 460,274 | 3.02 | 92,938 | 4.11 | 251,932 | 5.01 | 491,571 | 3.33 |
| Delaware | 5,127 | 2.07 | 1,801 | 2.72 | 3,326 | 1.84 | 6,106 | 2.47 | 2,430 | 3.67 | 3,676 | 2.03 |
| Maryland | 49,165 | 2.8 | 15,201 | 1.70 | 43,964 | 3.05 | 57,835 | 3.31 | 10,037 | 3.29 | 47,798 | 3.31 |
| Virginia | 60,903 | 2.66 | 16,653 | 2.43 | 44,250 | 2.75 | 92,672 | 4.04 | 43,979 | 6.42 | 48,693 | 3.03 |
| West Virginia | 12,267 | 1.78 | 9,619 | 2.29 | 2,648 | 0.99 | 27,212 | 3.95 | 23,561 | 5.60 | 3,651 | 1.36 |
| North Carolina | 68,228 | 2.71 | 34,101 | 2.78 | 34,127 | 2.64 | 97,639 | 3.88 | 58,899 | 4.81 | 38,740 | 3.00 |
| South Carolina | 48,670 | 3.87 | 24,332 | 4.42 | 24,338 | 3.44 | 62,417 | 4.96 | 35,034 | 6.36 | 27,383 | 3.87 |
| Georgia | 90,987 | 3.84 | 32,185 | 3.86 | 58,802 | 3.84 | 109,297 | 4.62 | 45,257 | 5.42 | 64,040 | 4.18 |
| Florida | 227,066 | 5.40 | 28,247 | 3.89 | 248,819 | 5.64 | 290,325 | 5.65 | 32,735 | 4.51 | 257,590 | 5.84 |
| East South Central | 183,199 | 3.49 | 85,316 | 3.79 | 97,883 | 3.23 | 266,446 | 5.01 | 156,600 | 6.75 | 109,846 | 3.61 |

*(continued)*

**TABLE 3.3** Continued

| Subregion and State | Overcrowded Housing Units | | | | | | Total Substandard Housing Units | | | | | |
|---|---|---|---|---|---|---|---|---|---|---|---|---|
| | Total | | Rural | | Urban | | Total | | Rural | | Urban | |
| | No. | Percentage[a] | No. | Percentage[a] | No. | Percentage[a] | No. | Percentage[a] | No. | Percentage[a] | No. | Percentage[a] |
| Kentucky | 34,118 | 2.47 | 17,395 | 2.73 | 16,723 | 2.26 | 64,233 | 4.66 | 45,154 | 7.07 | 19,079 | 2.57 |
| Tennessee | 48,393 | 2.61 | 17,476 | 2.50 | 30,917 | 2.68 | 69,648 | 3.76 | 35,195 | 5.04 | 34,453 | 2.98 |
| Alabama | 50,018 | 3.32 | 21,725 | 3.77 | 28,293 | 3.04 | 67,607 | 4.49 | 35,373 | 6.14 | 32,234 | 3.46 |
| Mississippi | 50,670 | 5.56 | 28,720 | 6.14 | 21,950 | 4.95 | 64,958 | 7.13 | 40,878 | 8.74 | 24,080 | 5.43 |
| South Central | 629,366 | 5.06 | 130,058 | 4.92 | 499,308 | 5.01 | 687,447 | 5.82 | 169,412 | 6.57 | 518,035 | 5.34 |
| Arkansas | 31,891 | 3.58 | 16,008 | 3.97 | 15,883 | 3.25 | 43,165 | 4.84 | 25,034 | 6.21 | 18,131 | 3.71 |
| Louisiana | 85,728 | 5.72 | 26,798 | 5.88 | 58,930 | 5.65 | 96,976 | 6.47 | 34,075 | 7.48 | 62,901 | 6.03 |
| Oklahoma | 37,891 | 3.14 | 12,689 | 3.43 | 25,202 | 3.01 | 44,187 | 3.66 | 16,987 | 4.59 | 27,200 | 3.25 |
| Texas | 473,856 | 7.81 | 74,563 | 6.39 | 399,293 | 8.14 | 503,119 | 8.29 | 93,316 | 8.00 | 409,803 | 8.36 |
| Total South region | 1,374,978 | 3.90 | 367,513 | 3.91 | 1,057,465 | 3.75 | 1,046,831 | 4.98 | 577,944 | 6.11 | 1,119,452 | 4.09 |

SOURCE: Housing Assistance (1994a).
a. Percentage of all occupied housing units.
b. District of Columbia excluded (data not available).

## Person-per-Room Criterion

The second criterion, crowding, was significantly higher in the region than was deficient plumbing (Tables 3.2 and 3.3). A unit is considered crowded if the persons-per-room ratio is greater than 1:1. More than 3.7 million units (3.9% of the region's total) were overcrowded. Rural units were slightly more crowded than urban units (3.91% vs. 3.75%). State-by-state comparisons reveal higher percentages of rural overcrowded units in Texas (6.39%), Mississippi (6.14%), and Louisiana (5.88%) than in other states.

## Total Substandard Units

The percentage of units lacking plumbing and the percentage over-crowded are not mutually exclusive. Units lacking plumbing or kitchen facilities may or may not be overcrowded. The total number of substandard units includes units deficient in one criterion or both criteria. Almost 5% (1,046,831) of the total units in the region were substandard. This includes approximately 6% of all rural units and 4% of all urban units. Across all divisions and states, the percentages of substandard urban units were lower than the percentages of substandard rural units. Mississippi, Texas, and Kentucky led all states in the percentage of rural substandard units, whereas the top three states in percentage of total substandard units were Texas, Mississippi, and Louisiana. In total, 577,944 rural units across the 18 states were substandard along the two measures of the Bureau of the Census. Although these criteria are inadequate indicators of housing quality, they are helpful in identifying concentrations of units deficient in size and plumbing facilities. Among the regional divisions, housing problems appear to be more severe in the East and West South Central divisions than in the South Atlantic.

In general, regional comparisons by division and state indicate that the magnitude of housing problems tends to be greater in the areas of the Lower Mississippi Delta (LMD). The Lower Mississippi Delta Development Commission (LMDDC) compared the 214 counties in the delta to the most underdeveloped countries in the world. Many historical economic, political, and social conditions bond the LMD counties in spirit and problems. The housing difficulties of the delta cannot be separated from this history. For this reason, housing conditions in the LMD are isolated for special attention.

# Housing in the Lower Mississippi Delta

Poverty and its attendant social ills are hallmarks of life in the 214 LMD counties of Mississippi, Arkansas, Louisiana, Tennessee, Kentucky, Missouri, and Illinois. Housing conditions of the region are merely symptoms of the larger socioeconomic and cultural traditions of life in the delta. Explanations abound as to the historical base and political climate that created such pervasive problems. Most acknowledge the role of the declining human resource needs of agriculture and its related industries. Less frequently cited are peculiar contributions of a biracial socioeconomic structure.

Although distinction is made between the LMD and the rest of the South, the history of the delta evolved within the context of the South as a whole. The southern caste system had its beginnings in the early colonial period and was centered around the region's dependency on labor-intensive agricultural products, principally cotton. Excessive labor demands led to the adoption of slavery, and those who succeeded in acquiring large holdings of land and labor developed into the aristocracy of the South. The agricultural way of life (the slave/master social stratification) dominated all phases of life, and for both economic and social reasons, the South was bent on maintaining a great void between the white and black races. The severity of housing problems in the region is not surprising given that the social, political, and economic agendas of the region were created, sanctioned, and nurtured on the economic exploitation and social isolation of a large percentage of the population. The rich soil of the delta was particularly suited to intense agricultural production and the "southern way of life."

The 214 counties of the LMD region contained approximately 3.4 million occupied housing units, up from the 2.2 million units reported in the 1980 census. Of these 3.4 million units, fewer than 2.1 million were owner occupied and more than 998,000 were renter occupied (Table 3.4). Housing quality data suggest that many of these units suffer serious structural and other deficiencies. Almost 57,000 units lacked complete plumbing, 143,000 units were overcrowded, and 650,000 were cost burdened. Deficient units were more prevalent in nonmetro counties than in metro counties, but cost-burdened units were found equally in both metro and nonmetro counties.

In terms of structural types, 10% of all occupied units were mobile homes (308,347 units) (Table 3.5). Of these units, 240,347 were owner

*(text continues on p. 43)*

**TABLE 3.4** Total Housing Units, Occupancy, and Tenure in the Lower Mississippi Delta

| Subregion and State | Total Units | Occupied Units | Percentage of Total | Vacant Units | Percentage of Total | Owner Occupied | Percentage of Occupied | Renter Occupied | Percentage of Occupied |
|---|---|---|---|---|---|---|---|---|---|
| Illinois[a] | 150,317 | 134,671 | 90.0 | 15,646 | 10.0 | 96,688 | 72.0 | 37,983 | 28.0 |
| Missouri[a] | 259,815 | 230,210 | 89.0 | 29,605 | 11.0 | 165,445 | 72.0 | 64,765 | 28.0 |
| East South Central[b] | 1,262,475 | 1,151,810 | 91.0 | 110,665 | 9.0 | 783,742 | 69.0 | 368,068 | 93.0 |
| Kentucky[a] | 196,441 | 177,598 | 90.0 | 18,843 | 10.0 | 128,662 | 72.0 | 48,936 | 28.0 |
| Tennessee | 549,390 | 505,096 | 92.0 | 44,294 | 8.0 | 326,102 | 65.0 | 178,994 | 35.0 |
| Metro | 341,867 | 316,604 | 93.0 | 25,263 | 7.0 | 189,870 | 60.0 | 126,734 | 40.0 |
| Nonmetro | 207,523 | 188,492 | 91.0 | 19,031 | 9.0 | 136,232 | 72.0 | 52,260 | 28.0 |
| Mississippi | 516,644 | 469,116 | 91.0 | 47,528 | 9.0 | 328,978 | 70.0 | 140,138 | 30.0 |
| Metro | 148,648 | 136,658 | 92.0 | 11,990 | 8.0 | 89,977 | 66.0 | 46,681 | 34.0 |
| Nonmetro | 367,996 | 332,458 | 90.0 | 35,538 | 10.0 | 239,001 | 72.0 | 93,457 | 28.0 |
| West South Central[b] | 1,737,159 | 1,532,630 | 89.0 | 204,529 | 11.0 | 1,004,857 | 66.5 | 527,773 | 67.0 |
| Arkansas | 560,040 | 501,983 | 90.0 | 58,057 | 10.0 | 338,957 | 68.0 | 163,026 | 32.0 |

*(continued)*

**TABLE 3.4** Continued

| Subregion and State | Total Units | Occupied Units | Percentage of Total | Vacant Units | Percentage of Total | Owner Occupied | Percentage of Occupied | Renter Occupied | Percentage of Occupied |
|---|---|---|---|---|---|---|---|---|---|
| Metro | 218,733 | 198,196 | 91.0 | 20,537 | 9.0 | 123,789 | 62.0 | 74,407 | 38.0 |
| Nonmetro | 341,307 | 303,787 | 89.0 | 37,520 | 11.0 | 215,168 | 71.0 | 88,619 | 29.0 |
| Louisiana | 1,177,119 | 1,030,647 | 88.0 | 146,472 | 12.0 | 665,900 | 65.0 | 364,747 | 35.0 |
| Metro | 700,917 | 613,207 | 87.0 | 87,710 | 13.0 | 359,259 | 59.0 | 253,948 | 41.0 |
| Nonmetro | 476,202 | 417,440 | 88.0 | 58,762 | 12.0 | 306,641 | 73.0 | 110,799 | 27.0 |
| Total Lower Mississippi Delta | 3,409,766 | 3,049,321 | 89.0 | 360,445 | 10.3 | 2,050,732 | 69.9 | 998,589 | 54.0 |
| Metro | 2,278,771 | 2,061,601 | 90.0 | 217,170 | 10.0 | 1,416,806 | 70.3 | 644,795 | 45.3 |
| Nonmetro | 2,227,039 | 2,032,700 | 91.0 | 194,339 | 9.0 | 1,362,295 | 67.0 | 670,405 | 48.5 |

SOURCE: Housing Assistance Council (1994b).

a. The Lower Mississippi Delta of Illinois, Kentucky, and Missouri contains only nonmetro counties.

b. Includes Lower Mississippi Delta states that are in the South.

**TABLE 3.5** Occupied Mobile Homes by Tenure in the Lower Mississippi Delta

| Subregion and State | Occupied Mobile Homes | Percentage of All Occupied Units | Owner-Occupied Mobile Homes | Percentage of All Owner-Occupied Units | Renter-Occupied Mobile Homes | Percentage of All Renter-Occupied Units |
|---|---|---|---|---|---|---|
| Illinois[a] | 18,693 | 14.0 | 13,107 | 14.0 | 5,586 | 15.0 |
| Missouri[a] | 29,314 | 13.0 | 21,692 | 13.0 | 7,622 | 12.0 |
| East South Central[b] | 106,614 | 10.3 | 83,316 | 16.5 | 62,105 | 7.7 |
| Kentucky[a] | 24,073 | 14.0 | 18,528 | 14.0 | 5,545 | 11.0 |
| Tennessee | 27,250 | 5.0 | 20,585 | 6.0 | 6,665 | 4.0 |
| Metro | 5,723 | 2.0 | 4,306 | 2.0 | 1,417 | 1.0 |
| Nonmetro | 21,527 | 11.0 | 16,279 | 12.0 | 5,248 | 10.0 |
| Mississippi | 55,291 | 12.0 | 44,203 | 13.0 | 11,088 | 8.0 |
| Metro | 7,790 | 6.0 | 6,305 | 7.0 | 1,485 | 3.0 |
| Nonmetro | 47,501 | 14.0 | 37,898 | 16.0 | 9,603 | 10.0 |
| West South Central[b] | 153,726 | 10.5 | 122,232 | 25.0 | 31,494 | 7.0 |
| Arkansas | 57,792 | 12.0 | 42,714 | 13.0 | 15,078 | 9.0 |
| Metro | 16,159 | 8.0 | 11,549 | 9.0 | 4,610 | 6.0 |
| Nonmetro | 41,633 | 14.0 | 31,165 | 14.0 | 10,468 | 12.0 |

(continued)

41

**TABLE 3.5** Continued

| Subregion and State | Occupied Mobile Homes | Percentage of All Occupied Units | Owner-Occupied Mobile Homes | Percentage of All Owner-Occupied Units | Renter-Occupied Mobile Homes | Percentage of All Renter-Occupied Units |
|---|---|---|---|---|---|---|
| Louisiana | 95,934 | 9.0 | 79,518 | 12.0 | 16,416 | 5.0 |
| Metro | 25,198 | 4.0 | 20,808 | 6.0 | 4,390 | 2.0 |
| Nonmetro | 70,736 | 17.0 | 58,710 | 19.0 | 12,026 | 11.0 |
| Total Lower Mississippi Delta | 308,347 | 10.0 | 240,347 | 12.0 | 68,000 | 7.0 |
| Metro | 54,870 | 4.0 | 42,968 | 6.0 | 11,902 | 2.0 |
| Nonmetro | 253,477 | 14.0 | 197,379 | 15.0 | 56,098 | 11.0 |

SOURCE: Housing Assistance Council (1994b).

a. The Lower Mississippi Delta of Illinois, Kentucky, and Missouri contains only nonmetro counties.

b. Includes Lower Mississippi Delta states that are in the South.

occupied (12% of all owner-occupied units). Mobile homes were 7% of all renter-occupied units. The percentages of occupied mobile homes appeared to be highest in Illinois, Kentucky, and Missouri; but these percentages are not comparable to percentages in other states because only nonmetro units were reported for these states. Across the LMD, mobile homes were considerably more prevalent in nonmetro areas. Mobile units comprised 17% of nonmetro owner-occupied units in the Louisiana counties of the LMD and 14% of the nonmetro owner-occupied units in the Arkansas and Mississippi counties of the LMD. The higher dependency on mobile homes in the nonmetro counties of the LMD is consistent with expectations based on other studies of mobile home use. Gruber, Shelton, and Hiatt (1988) found that the mobile home industry provides a low-cost housing alternative for low-income households. McCray (1992) noted that rural communities with higher percentages of substandard units tend to have higher percentages of mobile home units.

A report prepared for LMDDC compared housing conditions in the 214 target counties of the LMD with conditions in the remaining counties of the LMD states (McCray, Cotledge, Conley, & Watson, 1990). No significant differences were found between the aggregate mean scores of the 214 LMD counties and those of the other nontarget counties in the seven-state area on percentage of units lacking complete plumbing or overcrowded. However, some target county versus nontarget county differences were found within states. For example, in Mississippi, the mean number of units with overcrowding and deficient plumbing was significantly higher in the LMD counties.

The dollar value of all owner-occupied housing units differed significantly between the target and nontarget counties in the LMD region and in three of the seven LMD states (Illinois, Missouri, and Tennessee). Across the region, the percentages of owner-occupied units with a dollar value of $14,999 or less were significantly higher in the LMD counties, whereas the percentages of units valued between $15,000 and $34,999 were significantly lower in the LMD counties than in nontarget counties in the same states. No significant differences were found between the dollar value of owner-occupied units valued at $35,000 or above.

This study also analyzed differences in housing conditions by race of household head given racial dynamics of the LMD (McCray et al., 1990). For the region, the mean percentages of black-owned units valued at $4,999 or less were significantly higher in the LMD counties, and the mean percentages of black-owned units valued between $15,000 and $24,999 were significantly lower in the LMD counties than in other counties in the seven-state area. The only significant difference noted at the state level was in Louisiana, where the mean percentage of black-owned units in the LMD valued at $50,000 or above was significantly lower than in other counties of the state.

Median contract rents were significantly lower in the LMD region as a whole when compared with the median value aggregated across the non-LMD counties in the seven states. Median contract rents also were significantly lower in the LMD counties of three states (Arkansas, Illinois, and Missouri) when compared to the nontarget counties in the same states. No significant differences were found between the age of stock variable (measured by year in which structure was built) across the LMD counties in the seven-state area or in any of the individual states. Much of the housing stock of the LMD is aging.

Other characteristics indicative of housing problems in the 214 LMD counties were obtained from observations and experiences of life in the region. On-site visitations helped to visually document structural and environmental problems associated with the housing stock of the region. Throughout the LMD community, infrastructure was noticeably inferior in poor and minority neighborhoods as compared with that in other communities. In one Mississippi county, the air was filled with the stench of open sewage, water stood in almost every yard, and moisture problems associated with poor drainage were so severe that trees and wood framing on houses suffered from advanced stages of decay. Across the region, open ditches, dirt roads, and poor (if any) drainage systems are standard environmental conditions. Although many communities do have public sewage systems, some are ineffective and most are not available beyond short distances of the town boundaries.

The economics of slum housing in general, and in the LMD in particular, warrants notice. Slum housing is big business in almost any major city in any region of the country, but the business is much more lucrative in poor communities that lack regulatory devices and/or alternative living environments. LMD counties in the South continue to experience extensive housing and environmental problems. The work of LMDDC played a major role in documenting housing and other socioeconomic and cultural problems of the region. LMDDC's (1990) report recommends programming and policy development initiatives for revitalizing the region. Whether housing in the LMD will be improved as a result of LMDDC's effort remains to be seen.

## Challenges for the Region

Many citizens of the rural South live in better housing than at any time in history, but for many families, affordable housing is difficult to obtain and other housing problems remain. Housing affordability has worsened dramatically for all Americans since the 1970s. Rural housing deserves special attention because the dynamics of urban/rural housing deprivation often are misunderstood. Housing affordability is assumed to be the dominant urban problem, whereas housing quality is assumed to be the major rural issue. Such assumptions do not reflect the interface between housing quality and affordability, and they beg recognition of regional and urban/rural variations in income and housing characteristics. According to Housing Assistance Council data, 25.6% of the number of total housing units and 21.5% of all rural units in the South are cost burdened (Housing Assistance Council, 1994a). These percentages translate to 9.2 million and 2.4 million units, respectively.

Both housing affordability and housing quality are problems in many areas of the rural South, but affordability is more difficult to comprehend because of an inconsistent definition and a lack of clear visibility. In the building industry, affordable housing means modest, no-frills units that can be constructed for a specified cost. The cost is determined by economic conditions and sociodemographic characteristics of the market area. In the rural South, a newly constructed house in the $40,000 to $50,000 range may be marketed as "affordable." In other regions of the country, the price tag for an "affordable house" may exceed $100,000. In essence, the building industry equates affordable with lower than average construction costs. This definition is purely economic and lacks reference to a specific household or consumption unit.

From another perspective, affordable implies a match between consumer income and housing cost. People live in low-quality housing when they cannot afford to pay market prices for standard units, when there is a limited supply of low-cost standard units, and/or when housing subsidies are not available. When the term "affordable housing" is tied to a household unit, it is neither expensive nor inexpensive; it simply matches the household's ability to pay (McCray, 1989).

Typically, housing affordability for a specific market area is calculated as a ratio of some average income figure. Such measures fail to recognize that low-quality housing is less costly because it fails to meet prescribed standards of decency and/or safety. In a reasonably efficient housing market, areas with low incomes should have low-cost housing that provides decent and safe living environments. Housing prices and ability to pay should balance. The result of an imbalance between low-cost housing and low incomes is the overconsumption of low-quality housing units or a disproportionate number of families with excessive housing costs as compared with other family expenditures. Typically, the rural response has been the overconsumption of low-quality units.

But what is the price that rural communities pay for this imbalance? Both housing quality and affordability issues negatively influence the "attachment to place" of ill-housed residents. As a result, strong emotional bonds to community and family often are diminished. According to Marans and Wellman (1978), place of residence has a direct bearing on quality of life. Economic well-being is germane to the overall assessment of quality of life, with neighborhood and housing factors weighing heavily in the overall assessment. Small rural towns often experience deficits in the quality or level of community services, creating dissatisfaction with small-town life.

The availability of federal funds directly influences the development of subsidized housing for low- and moderate-income families in the rural South. A reduction in funding for housing programs can exacerbate housing conditions. Some problems with affordability also can be attributed to the preference of the American public for the single-family, detached, conventionally built home—the most expensive form of housing. However, its use often is encouraged through the local codes and regulations as well as lending, building, and marketing practices.

Several alternative housing approaches have the potential to lower initial and/or maintenance costs of housing. Although some affordable housing options (e.g., mobile homes) have met with acceptance, other more innovative options have experienced limited adoption.

In recent years, local communities have begun to recognize the importance of housing and neighborhood characteristics in enhancing the retail and industrial bases of small communities. Earhart, Weber, and McCray (1994) found that resident satisfaction with housing and neighborhood characteristics has implications for community development. They found a significant negative association between housing satisfaction and desire to relocate among rural baby boomers. Elderly residents of rural communities create another challenge. It has been found repeatedly that older persons are more satisfied with their housing. Findings in the Earhart et al. (1994) study are no different. A primary policy concern is assisting elderly residents in their desire to "age in place." This is particularly critical in rural areas, where the housing of the elderly tends to be older and in poorer structural condition.

Housing units are tied to households, but housing issues transcend household boundaries. Constraints to obtaining quality housing must be viewed as a function not only of the family but also of the community. Housing problems link families and communities in systemic social networks that require multifaceted responses and partnerships for resolution.

A number of housing and environmental problems are associated with a failure of local communities to regulate building construction, occupancy standards, and land use. Although some literature suggests that local regulations may increase the cost of housing, such strategies also protect community residents from unsafe environments. Enforcement of building and/or occupancy standards is dependent on an adequate supply of standard housing units, an elusive luxury for many rural communities in the South.

Success in the delivery of adequate and affordable housing in rural areas, especially housing incorporating new approaches, is a complex process of interactions between and among consumers, intermediaries (lenders, builders, realtors, and others involved in the housing process), elected and appointed officials, local and state regulators, and individual leaders concerned about housing. The attitudes and interests of various market segments are influential in determining housing availability in rural communities.

Although 1990 data show marked improvements in the quality and quantity of rural housing in the South over the previous 10 years, the "national housing goal"—a decent home and suitable living environment for every American family—is far from being fulfilled. On the other hand, the status of rural housing in the South is improving relative to other regions of the country. Such improvements might be the result of the highly touted "rural renaissance" of the 1970s, changing social and political dynamics of the region, or both. Nonetheless, with a few exceptions (the LMD and Appalachian counties), the rural South has experienced positive change in the relative status of its housing stock.

# Note

1. The Bureau of the Census's "South" consists of Delaware, Maryland, Virginia, West Virginia, North Carolina, South Carolina, Georgia, and Florida in the South Atlantic subregion; Kentucky, Tennessee, Alabama, and Mississippi in the East South Central subregion; and Arkansas, Louisiana, Oklahoma, and Texas in the West South Central subregion.

# Chapter 4

# The Border *Colonias:* A Framework for Change

*Zixta Q. Martínez*
*Charles Kamasaki*
*Surabhi Dabir*

The term *colonia* has its origins in the Spanish word for "community" or "neighborhood." In the United States, it describes the hundreds of quasi-rural communities located along the U.S.-Mexico border and characterized by extreme poverty and severely substandard living conditions. The border colonias are defined primarily by what they lack, such as safe drinking water, water and wastewater systems, paved streets, and standard mortgage financing. Over the years, a number of studies and needs assessments have documented the specific needs of colonias. These reports have highlighted the grim living conditions in these communities, chronicled their infrastructural constraints, developed plans for improvements, and prepared cost estimates and recommendations aimed at the government.[1] Rather than duplicate these efforts, this chapter presents a framework that recognizes that the colonias are not an isolated or random phenomenon but rather the result of larger systemic forces. It also argues that the problems in the colonias require and are amenable to solution through firm, immediate policy and program intervention.

Colonias will not soon simply disappear. To the contrary, historical trends, confirmed by recent demographic data, suggest that the colonia phenomenon, with all the attendant social and economic problems, will continue to proliferate absent immediate and sustained action. Any long-term, cost-effective approach to address the problems in the colonias must recognize that, left unattended, the scope, degree, and costs of remedying problems associated with colonias will escalate and ultimately will affect the economic and social well-being of all the border states.

## Background and History

The 1990 National Affordable Housing Act (NAHA) created a federal definition for the colonias: an "identifiable community" in Arizona, California, New Mexico, or Texas within 150 miles of the U.S.-Mexico border, lacking decent water and sewage systems and decent housing, and in existence as a colonia before November 28, 1990. Access to the program, funded by the U.S. Department of Housing and Urban Development (HUD), is tied to this definition of the colonias. Prior to this, a 1983 border environmental agreement between Mexico and the United States (La Paz Agreement) defined the "border region" as a zone within 100 kilometers (approximately 62 miles) on either side of the political boundary. Colonias falling within this 62-mile zone were eligible for Environmental Protection Agency (EPA) programs. Similarly, individual state and county programs often stipulate their own eligibility requirements for colonias. In fact, access to most colonia assistance programs and funding is tied to a specific definition, posing a challenge to communities seeking aid from multiple funding sources. More important, none of these official definitions captures the human dimensions of the problems faced by colonia residents.

Historically, some colonia developments began as small communities of farm laborers employed by a single rancher or farmer. Others originated as town sites established by realtors in the early 1900s. The majority of colonias, however, emerged in the 1950s as land developers and speculators discovered a large market of aspiring home buyers who could not afford homes in cities or access conventional mortgage financing.

For the past several decades, the U.S.-Mexico border region has experienced severe developmental pressures due to industrialization, immigration, and population growth. Infrastructure to meet environmental, health, housing, transportation, and other needs has not kept pace with this development. As a result, colonias have continued to proliferate. Furthermore, despite state legislation intended to prevent such growth, legal loopholes and weak enforcement of laws continue to allow developers to sell land under a "contract for sale"[2] to unsuspecting residents eager to own a piece of the American dream.

Although many describe the development of colonias as simply a land use phenomenon, broader social and economic forces have played a significant role in their rise. The principal forces that drive individuals to these colonias include high poverty rates in the border region and a severe lack of affordable housing. People choose to live in colonias primarily because they cannot afford to live elsewhere.

Moreover, the colonia phenomenon has been driven in part by the complexities of the domestic and international agricultural labor markets. The colonias of the Lower Rio Grande Valley serve as the principal "home base" for many migrant farm laborers. Similarly, the California communities that resemble colonias, particularly in the Imperial and Salinas valleys, house many of the families who make up the California migrant stream. Farmworkers suffer from a virtual absence of effective labor law coverage and enforcement, the transitory and seasonal nature of farmwork, and the traditional practice of reliance on immigrant workers from Mexico to ensure an oversupply of agricultural labor. In addition, the colonia phenomenon effectively subsidizes agriculture throughout the United States and helps to maintain the country's traditionally low consumer prices for fresh produce.[3] Although colonias may be isolated from and often invisible to nearby town and city residents, they represent an integral element of the U.S. economy.

## Current Status

Typical descriptions of colonias graphically depict the substandard living conditions faced by residents, with frequent comparisons to squatter settlements in developing countries. Colonia communities often are concentrated pockets of poverty, characterized by dirt roads without adequate drainage control and substandard, overcrowded housing. This housing often is constructed out of "found materials" such as cardboard, wood palettes, and corrugated metal sheets. It also is not uncommon to find people living in abandoned buses or tar paper shacks. Colonia residents often lack access to potable drinking water, use illegal cesspools or septic tanks, and/or dispose of waste directly into open trenches. The use of dangerous Mexican butane tanks for heating and cooking also is commonplace, and residents suffer from a high incidence of preventable diseases.[4]

These descriptions, although accurate, fail to capture fully the very diverse living conditions experienced in colonias. Colonias vary widely in the limits imposed by local economics and politics and by their infrastructure needs, which differ with respect to age, geography, and proximity to cities. Because of this diversity, and because colonia residents tend to be politically disenfranchised, they remain largely unheard in national debates. However, colonia residents do represent a significantly large population.

In Texas alone, nearly 1.7 million people lived along the border region at the time of the 1990 census. The Texas Water Development Board (TWDB) conducted a survey of colonias in Texas in 1992, estimating that 279,863 residents lived in 1,193 colonias. In 1995, TWDB updated this figure to 340,000 estimated residents in 1,436 colonias. Nearly 60% of these residents live in the four-county region known as the Lower Rio Grande Valley, which consists of Hidalgo, Willacy, Starr, and Cameron counties. These border counties consistently rank among the poorest in the country.

A 1990 General Accounting Office (GAO) report found that New Mexico identified colonias only in Dona Ana County. According to GAO, there were 15 colonias in Doña Ana County alone, with a total population of 14,600. However, as of November 1994, as many as 60 colonias had been identified in four counties: Doña Ana, Grant, Luna, and Hidalgo.[5] A representative of the local government division in New Mexico estimated that 80,000 to 100,000 individuals resided in New Mexico's colonias.[6] These discrepancies almost certainly represent both definitional differences and the actual growth of colonias in New Mexico.[7]

California and Arizona both have lagged behind Texas and New Mexico in recognizing the presence of colonias. Some California state and San Diego County officials have in the past asserted that colonias simply do not exist in California. However, a 1987 Congressional Research Service (CRS) report identified a number of colonias on the outskirts of San Diego and in Imperial County. CRS reported a population of 25,000 living in the colonias of San Diego County and 11,500 in the colonias of Imperial County. A more recent study by Rochin and Castillo indicates that many of California's rural communities with a high Latino concentration are characterized by conditions common to colonias. Most of these communities, however, apparently do not meet the legislative definition of colonias, in part because they are located outside the border area.[8]

The 1987 CRS report also identified a total of 50 to 55 Arizona colonias in Yuma, Pima, Santa Cruz, and Cochise counties as well as in the interior counties of the state, along Interstate 10, and outside the cities of Tucson, Casa Grande, and Phoenix. As of November 1994, Arizona officially recognized colonias in Yuma and Cochise counties and outside the cities of Douglas and San Luis and the town of Welton. The estimated population of these colonias is 9,025.[9]

## The Policy Context

### Colonia Assistance Programs

A number of state and federal agencies provide assistance to colonias. Under Section 916 of the NAHA, the four border states are required to set aside up to 10% of their Community Development Block Grant (CDBG) funds for use in the colonias. Although Texas and New Mexico have consistently set aside the full 10% of their CDBG dollars for use in the colonias, Arizona's and California's reluctance to formally recognize colonias at first prevented the provision of badly needed grant funds for the colonias.

In addition, each of the border states except Arizona established a different mechanism for distributing funding beginning in fiscal year 1991. Texas restricted its application competition to 70 border counties. In 1991, Texas received only seven applications for a total request of $2,391,146 out of the state's allocation of $5,432,800. New Mexico restricted its application competition to four counties and five municipalities. Four applications were received, and three received funding. Although California did not formally recognize any colonias, migrant camps in Imperial County met the statutory definition and received $160,000. California did not establish a special application competition because it determined that only a few communities could meet the definition of a colonia.[10]

All four states have now begun to use the federally mandated CDBG funds set aside for the colonias. Arizona has set aside a full 10% of its CDBG funds for its colonias since fiscal year 1994, and California has set aside between 2% and 5% of its CDBG total for its colonias since fiscal year 1992. The set aside legislation itself was permanently extended in 1997.

Several existing federal programs have long been used to assist colonia residents. These include the U.S. Department of Agriculture's grants and

loans for water/wastewater and housing, EPA wastewater construction grants
and loans, and various housing programs of HUD. Few of these, however,
explicitly target colonias, and all face formidable political and other obstacles in securing significant funds. Several other national efforts at mobilizing
expertise and resources to address the needs of the colonias also have met
with failure.[11]

State resources to improve conditions in the colonias are even more
constrained. Only in Texas has there been a significant state-level effort to
help colonias.[12] This effort is impressive in scope and vision but continues
to experience implementation problems.

### Barriers to Effective Remedial
### Programs and Policies

Like other federal development and assistance efforts targeted to a
predominantly minority, low-income population, colonia development
strategies face significant obstacles. These include insufficient resources,
lack of consensus on appropriate development approaches, widespread skepticism and even cynicism about government-sponsored community development, and the phenomenon known as "compassion fatigue." This is the
unwillingness of some segments of the American public to support major
antipoverty programs, particularly when the problems they address seem
intractable. These difficulties are magnified by the extraordinary scope and
degree of problems faced by residents of colonias.

In addition, several other factors hinder the development of effective
programs and policies to address the social, physical, and economic problems faced by colonia residents. These barriers may be grouped into four
categories: legal and physical constraints, local political considerations,
national program and political limitations, and the absence of an effective
advocacy or service delivery infrastructure.

Physical and legal isolation makes very difficult any changes in programs and policies to help in the colonias. Consider the effects of physical
isolation. The provision of basic water, sewer, and paving services is considerably more costly for colonias compared with inner city communities,
especially for those that are not located adjacent to cities. This is true because
of inherent diseconomies of scale associated with small community size.
Construction of treatment plants for such small communities also is generally

not economically feasible. Similarly, the extension of water distribution and wastewater collection lines from existing treatment facilities to remote geographical locations tends to be prohibitively expensive.

In addition, colonias often fall outside the jurisdiction of cities. Counties often have limited legal power to incur debt needed to finance infrastructure improvements. Cities are hesitant to either extend services to or annex colonias adjacent to their corporate limits because these poorer communities offer a negligible addition to the tax base. Moreover, given their extraordinarily low incomes, most colonia residents cannot afford city taxes and user fees.

These problems are compounded by state and local political considerations. Because of their dispersed nature, even in areas of their highest concentration such as Hidalgo County in Texas, colonia residents make up only a small fraction of the overall county population, which is concentrated in the cities. In addition, like other predominantly low-income and minority communities, colonia residents have little political power to compete for scarce resources.

National political considerations also complicate the creation of programs and policies targeted to colonias, which are located in a few counties in only four border states. Colonias tend to fall outside of classic rural versus urban definitions, posing a challenge with regard to program eligibility requirements. Furthermore, there is a widespread perception that most colonia residents are immigrants. At a time of almost unprecedented "immigrant bashing," prospects for new national development efforts targeting this population appear dim.

Finally, in addition to the absence of physical infrastructure, colonias appear to suffer from a lack of adequate "development infrastructure." In the late 1960s and early 1970s, the Lower Rio Grande Valley in general, and colonias in particular, had a strong network of community organizations that served as both advocates and service providers. This coincided with a lack of Latino elected officials in the area and with the height of the War on Poverty and the Hispanic civil rights movement. Budget cuts and consolidation of federal programs into block grants to states and localities severely reduced the funding base of these community groups.[13] Subsequently, there was an increase in local Latino elected officials.[15] Paradoxically, this did not result in a corresponding increase in resources available to local colonia groups. As a result, even when development and service delivery opportunities become available, they frequently are underused. Moreover, the re-

Moreover, the resources and expertise that community-based groups typically bring to urban settings are largely unavailable in colonias.[14]

# A Framework for Colonia Development Efforts

## Why Bother?

As the nation enters a period of diminishing resources, increased skepticism and cynicism about government, and widespread compassion fatigue, one might legitimately question whether any substantial federal program to help the colonias is either justified or achievable. However, such an effort is not only justified but also essential on at least four independent levels.

First, on an economic level, immediate action is the most responsible and cost-effective policy option. Despite efforts to the contrary, the number and size of colonias, and thus the magnitude of the problem and the cost of remedial action, continue to grow. In fact, the costs of ignoring the problem, in direct public assistance as well as through "hidden" expenses such as health care expenditures, are significant. The well-being of the border states clearly is hampered by the presence of such pockets of poverty, which impede economic development.

Second, on a social level, no society can tolerate the existence of a large group of "second-class" citizens living in terrible deprivation. If a society should be judged by how it treats its disadvantaged, then the extent to which the residents of colonias are afforded the basic necessities of life is of fundamental importance. Increased public investment along the border generally, and in colonias particularly, is justified on simple equity grounds. A number of studies strongly suggest that the residents of these areas receive a disproportionately low "rate of return" on their tax dollars in terms of public expenditures.[16]

Third, on a pragmatic level, the provision of basic water and sewer systems, paved streets, and decent housing to virtually all colonia residents ought to be an achievable public policy objective. Although cost estimates vary greatly, significant federal resources are available from HUD, the Department of Agriculture, and other agencies. When combined with other public and private resources, a very good start on eliminating the problems of colonias can be made with current funding levels. In comparison to many other social policy aims, significant change is well within current levels of funding.

Finally, in addition to improved physical infrastructure, any colonia development efforts would provide substantial ancillary benefits. Such efforts would generate substantial economic activity for construction firms and related service industries in a highly impoverished region. At least some of the new jobs created could provide employment opportunities for colonia residents themselves.

Furthermore, a comprehensive assistance effort would provide a unique opportunity for the development and strengthening of community-based social services, development, and advocacy infrastructure within the colonias. The long-term benefits and effects associated with improvements to the colonias certainly will go beyond the relatively modest initial investments.

## Recommendations

A number of public and private agencies and coalitions active along the U.S.-Mexico border have developed detailed program and policy recommendations for addressing the problems in the colonias. Although these recommendations vary in some particulars, most agree that a coordinated public and private effort by local, state, and national organizations is key to improving conditions. To be successful, a colonia assistance initiative must involve the broadest range of participants and resources. The recommendations presented here are intended to support existing efforts in the colonias and suggest a broad framework for a comprehensive assistance strategy for the border.

*Comprehensiveness: Colonia assistance efforts should address all three basic needs—water and wastewater systems, housing and economic development, and social services.* Some advocates argue that the provision of water and other physical infrastructure should have the highest priority because such systems are essential to protect the public health and frequently are a prerequisite for other development. Others argue that, without economic development and social services, colonia residents will not be able to maintain the physical infrastructure given their small tax base and inherent diseconomies of scale. This tension is a direct result of limited access to scarce resources. A comprehensive assistance effort would provide a unique opportunity to strengthen community-based social services, development, and advocacy in the colonias.

*Flexibility and Innovation: To the maximum extent possible, colonia assistance efforts should devise creative approaches to meeting the needs of the residents.* One way in which to reduce both initial capital and long-term maintenance costs of water and wastewater systems is through the use of innovative technologies. Such "low-tech" technologies are available with respect to housing rehabilitation and construction but may be precluded by local building codes or federal standards. There clearly is a need for such regulations, but it is ironic that colonia residents could be deprived of safe housing and sanitary water systems by the same laws that have failed to protect them from health and environmental hazards in the first place.

In addition, any colonia assistance effort should take into account the patterns of employment and income of colonia residents. Most colonia residents have a very strong desire for homeownership. They also have a deeply rooted ethic of "sweat equity," adding to and improving their homes as their savings and income permit. It should be possible to develop creative financing mechanisms and innovative programs that accommodate these characteristics.

*Early Success: To the maximum extent feasible, and consistent with an appropriate allocation of limited resources, initial development efforts must succeed.* Given the political vulnerability of funding targeted to colonias, assistance initiatives must show that program objectives can be met in a cost-effective manner. For example, colonias adjacent to cities might be targeted for assistance first through the provision of financial incentives to the cities that would promote annexation or the provision of basic services. Colonias located in particularly remote regions might be targeted for low-cost, low-tech water and wastewater systems. Social services, housing, and economic development might be emphasized for those colonias that have basic physical infrastructure.

*Collaboration: Colonia assistance efforts should promote cooperation and participation of a broad range of stakeholders.* Federal, state, county, and city governments, as well as community-based organizing, social service, and development groups and colonia residents themselves, all have important roles to play and should be included in solving problems. Collaboration between public and private entities also should be encouraged. This is especially true at a time of uncertain funding and close scrutiny of all development efforts.

Careful but vigorous implementation of a comprehensive, flexible, inclusive, and strategically designed colonia assistance effort can simultaneously improve the daily lives of people, promote the public health and economic well-being of a historically impoverished and neglected region, and bring the U.S. society closer to ideals of justice and fairness. Unlike so many of the social ills now affecting the United States, turning the colonias around can be accomplished at a comparatively modest cost. It is rare that policymakers and program operators are afforded such a clear opportunity to both "do well" and "do good." It is an opportunity that should not be missed.

## Notes

1. See Border Low Income Housing Coalition (1993), Governor's Border Working Group (1993), Office of the State Auditor of Texas (1993), and Texas Department of Housing and Community Affairs (1993).

2. Under this type of "contract for sale" arrangement, the land title stayed with the developer until the price of the property was paid in full, and repossessions by developers as a consequence of one missed payment were not uncommon. Plots frequently were purchased with the explicit, but not legally enforceable, understanding that the developers would provide necessary improvements such as water, sewers, drainage, and paved streets. In fact, such systems rarely were installed.

3. See, for example, Commission on Agriculture Workers (1992).

4. Detailed descriptions of conditions in the Texas colonias can be obtained from *Colonias Housing and Infrastructure,* a three-volume series published by the LBJ School of Public Affairs under contract with HUD (Chapa & Eaton, 1997). A synthesis of 1990 census data on all border counties with the presence of colonias is included in Housing Assistance Council (1994b).

5. Conversation with Clyde Hudson, Farmers Home Administration in New Mexico, November 1, 1994.

6. Conversation with Lynn Goldstein, November 1994.

7. This assertion is consistent with a 1987 Congressional Research Service (CRS) report that identified colonias in the same four counties of New Mexico, suggesting that the reported growth in number from 15 to 60 represents both actual growth and definitional differences.

8. See Rochin and Castillo (1993).

9. Conversation with Rivko Knox, Community Development Block Grant program manager, Arizona Department of Commerce, November 1994.

10. See Communications Group (1993).

11. One example is the Colonias Assistance Program proposed by HUD Secretary Henry Cisneros in 1994. Although funds for this program were appropriated for fiscal year 1995, the authorizing legislation was not enacted. Since then, this initiative has not been revived. Similarly,

the Interagency Task Force on the Colonias convened by HUD during Cisneros's term is now defunct.

12. The Texas state legislature has formed a Senate Committee on International Relations, Trade, and Technology. The committee has convened several hearings on the colonias and passed several pieces of legislation to address problems in the colonias. The legislation provides basic consumer protection to people purchasing land in border counties, directs the Texas Department of Housing and Community Affairs to set up five colonia self-help housing centers, sets aside funds for colonia-related purposes, reserves $40 million in tax-exempt mortgage bonds for the colonias, and halts land sales in the colonias where lots lack water and sewer facilities that meet state standards under the Economically Distressed Areas Program subdivision regulation. Other border states have not shown a matching degree of concern or interest.

13. Nationally, for example, more than 40 of the National Council of La Raza's (NCLR) affiliates went out of business in the early 1980s, principally as a result of budget cuts and block grants. Perhaps 10% of these groups served colonias. This community group infrastructure has yet to be fully replaced. Most observers would argue that the emergence of the Industrial Areas Foundation (IAF)-affiliated Valley Interfaith, EPISO, and other community organizations that placed a high priority on colonia issues increased the overall political advocacy power of colonia residents. It did not, however, replace the technical and organizational service delivery and development capacity of community groups, and in some respects it diminished it further. This is true because the IAF groups tend to view community groups as competitors and in effect promote the primacy of local elected officials in return for the enactment of favorable policies.

14. It should be noted, however, that a number of organizations including Texas Rural Legal Aid, the Neighborhood Reinvestment Corporation, NCLR, and Housing Assistance Council have been actively assisting in the development of new, emerging development and service delivery groups in colonias.

15. The growth in Hispanic elected officials in border states in the late 1970s and early 1980s, stimulated in large part by the 1975 Voting Rights Act, negatively affected the status of local Latino community-based groups in some respects. Paradoxically, it appeared to have resulted in stronger local government control over the level of resources available to serve colonias and might even have resulted in a net reduction of resources flowing to colonia-serving organizations. It diminished both the political "leverage" that community groups had with what previously had been Anglo-dominated elected institutions, and it increased the ability of local elected officials to channel resources to divisions of local government, sometimes at the expense of community groups.

16. Texas border counties were among those found to be negatively affected by unconstitutional (under the Texas State Constitution) disparities in school finance patterns in the *Edgewood Independent School District v. Kirby* case regarding elementary and secondary school funding. In the *Edgewood* case, poor, predominantly minority school districts throughout Texas successfully argued that the state's school finance system failed to equitably serve such districts. A group of plaintiffs and several South Texas colleges and universities made similar arguments with respect to higher education funding patterns in *LULAC v. Richards*. Although the plaintiffs in this case were unsuccessful, the factual claim that South Texas received disproportionately less funding than the rest of the state was essentially unrefuted. Similar disparities have been unearthed with respect to highway funds, social services, mental health services, and other state-funded programs in Texas as a result of a "Court of Inquiry" procedure initiated by El Paso State District Court Judge Eduardo Marquez. See, for example, "El Paso Comes Up Short in Public Aid" (1994), "Highway Boss Can't Explain Disparities" (1994), "Border Benefits" (1994), and "Unfair Funding Spurs Inquiries" (1994).

# Chapter 5

# Affordable Housing
# in the Rural Midwest

*Janet A. Krofta*
*Sue R. Crull*
*Christine C. Cook*

The Midwest is known as the country's "Heartland."[1] Geographically in the north central part of the continental United States, it also has been viewed as possessing a core of solid rural values of hard work and strong families. The Midwest is the locus of an imagined rural life that is peaceful, harmonious, and healthy. It was the logical site of the "field of dreams" celebrating heroes of the great American sport of baseball. As one observer has noted, "Midwesterners feel themselves and their way of life to be distinct from other people and other lifestyles, and they consider the quality of their lives to be second to none" (Dunn, 1987, p. 50).

The Midwest is valued throughout the world for its abundant agriculture, made possible by millions of acres of fertile farmland. Industrialization of the region developed out of the need for machines to produce, harvest, process, and transport its agricultural products. However, the growth of huge farm surpluses, the switch from an industrial economy to an information economy, and the many farm failings in the early 1980s have changed the rural midwestern landscape. These changes in the regional economy, along with out-migration of young households and an aging of the population and housing stock, now inform the current housing needs and conditions of midwestern households.

This chapter explores the housing reality for people who live in the rural Heartland. Both population and housing characteristics are examined, with particular attention to elderly, female-headed, and minority households. Contrasting images are presented. For many households in the Midwest, housing is adequate and income is ample; midwestern living represents the "American dream." For other households, however, the condition and cost of housing, coupled with shrinking employment opportunities, present a more dismal future.

## Population and Housing Characteristics of Nonmetropolitan Midwest Households

Population and housing characteristics are inextricably linked. To understand the supply of and demand for housing in the rural or nonmetropolitan (also referred to as "nonmetro") Midwest, it is essential to examine the demographic characteristics that distinguish the region from metropolitan ("metro") areas and the U.S. population as a whole. The age of a household, familial status (e.g., married, widowed, single parent), number of children, race/ethnicity, income, and other factors are strongly associated with the ability of households to access and occupy decent, affordable, and appropriate housing.

In general, nonmetro households in the Midwest, indeed around the nation, are different from metro households in that they are more likely to be white families and less likely to be young single individuals or single parents (Morris & Winter, 1982, p. 196). Rural households also are more likely than their urban counterparts to be headed by elderly persons.

Table 5.1 compares households in the nonmetropolitan Midwest with all U.S. households in 1995.[2] It shows that the average midwesterner was likely to be older and less likely to live in a household with children present. The smaller number of young people in the Midwest is due to the out-migration of adults in their childbearing years and to an "aging in place" of the midwestern population. The proportion of households headed by persons age 65 years or over was considerably higher in the Midwest than in the United States overall. About half of all elderly households in the Midwest were married couples; most of the remaining ones were widowed women living

**TABLE 5.1.** Population and Housing Characteristics of Nonmetropolitan Midwestern Households Compared With U.S. Total Households (percentages unless otherwise indicated)

| | Midwest Nonmetropolitan | U.S. Households |
|---|---|---|
| Population characteristics | | |
| Age (mean years) | 49 | 46 |
| Household size (mean) | 2.2 | 2.3 |
| Married | 57 | 52 |
| Children under 6 years of age present | 14 | 18 |
| Children under 18 years of age present | 36 | 38 |
| Elderly households (age 65 years or over) | 28 | 21 |
| Female-headed households | 7 | 10 |
| Minority households | 4 | 23 |
| Monthly income (mean dollars) | 2,304 | 2,618 |
| Poverty | 16 | 15 |
| Receiving welfare or supplemental security income | 5 | 6 |
| Receiving housing assistance | 18 | 15 |
| Housing characteristics | | |
| Year house was built (median) | 1958 | 1968 |
| Purchase price (median dollars) | 33,189 | 51,305 |
| Home value (median dollars) | 63,740 | 92,507 |
| Housing expense per month (mean dollars) | 370 | 543 |
| Housing costs more than 35% of income | 19 | 25 |
| Housing costs more than 50% of income | 10 | 14 |
| Homeowner | 75 | 65 |
| Single-family dwelling | 79 | 68 |
| Mobile home | 8 | 6 |

SOURCE: U.S. Bureau of the Census and U.S. Department of Housing and Urban Development (1997).

alone (U.S. Bureau of the Census & U.S. Department of Housing and Urban Development [HUD], 1997). Although many of these women may have family and friends nearby, there is increasing danger that, as they become frailer, the availability of caregivers, proper medical attention, and appropriate housing will diminish.

In contrast to the number of elderly households, the proportions of female-headed and minority households were smaller in the nonmetropolitan Midwest[3] than in the United States overall. Minority households were especially underrepresented in the region, accounting for about 4% of all households compared with about one quarter nationally.

In terms of income, the average nonmetro midwesterner earned $300 less per month than did the average U.S. household. The proportions living in poverty were similar in midwestern households and in U.S. households overall. Nonmetro midwesterners were, however, slightly more likely than their U.S. counterparts to receive housing assistance.[4]

Table 5.1 also compares housing units in the rural Midwest with units in the nation as a whole in 1995. In general, the Midwest had higher percentages of single-family detached dwellings (79% vs. 68%), mobile homes (8% vs. 6%),[5] and seasonal units such as vacation homes. Although the image of a house in the region may be that of a farm dwelling, only about 9% of all the owner-occupied housing was located on farms (U.S. Bureau of the Census & HUD, 1997). Alone or in clusters, houses often are ranch-style or two-story wood frame buildings with basements and pitched roofs. Seasonal homes may be simple cabins or luxurious dwellings.

The median value of homes and mean monthly housing expense both were lower in nonmetropolitan areas of the Midwest than in the United States as a whole. This might explain, at least in part, why the region had a higher rate of homeownership (75% vs. 65%). In 1995, median home value in the rural Midwest was $63,740 compared with $92,507 nationwide. The mean monthly housing expense was $370 compared with $543 nationwide. Nearly 1 in 10 households spent more than half of its income on housing. For many of these households, there can be little left over for other living expenses.

Whether on farms or in rural communities, midwestern single-family houses tend to be old and in need of modernization (U.S. Bureau of the Census & HUD, 1997). The Midwest was second only to the Northeast in the proportion of housing more than five decades old (25% vs. 33%) (U.S. Bureau of the Census, 1990). Only about 14% of the housing stock was built between 1980 and 1990 compared with 21% nationwide.

Most of these rural homes are not self-sufficient. Just over half of the units surveyed in the 1995 American Housing Survey (AHS) used piped gas from a central system as their main heating fuel compared with only 21% nationally. Two thirds reported that their water supply was from a public system or water company, and nearly two thirds were on public sewer lines.

As might be expected with an aging population, half of owner-occupied units were free and clear of mortgage debt (U.S. Bureau of the Census & HUD, 1997). For those with one or more mortgages, however, the median year of mortgage origination was 1991. The overwhelming majority of mortgages were conventionally financed without benefit of government direct, insured, or guaranteed mortgage programs. Fewer than 5% used Federal Housing Administration (FHA) mortgages, 4% used Veterans Administration (VA) mortgages, and 3% used Rural Housing Service (RHS, formerly Farmers Home Administration [FmHA]) mortgages. State and local programs to lower the cost of financing were used nearly as often as were FHA, VA, and RHS mortgages combined.

The mortgage credit market also has changed over time. Some communities have lost local banks along with other businesses during times of economic decline, and access to mortgage credit has decreased. Borrowers must, then, look to other types of mortgage lenders. In other places, bank mergers have provided access to larger lending pools and greater lending sophistication. The Midwest is not unique in this. Access to credit is more a function of population density, economic activity, and personal income than of regional location.

# Housing and Housing-Related Concerns in the Midwest

In this section, four important issues of concern to rural households in the Midwest are discussed: (a) the housing problems of special populations (the elderly, female-headed households, and minorities), (b) the unique characteristics of vacation and retirement housing and communities, (c) the changing quality of the environment, and (d) the changing regional economy and its effect on housing. Although some of the concerns described in this discussion are not uniquely midwestern, they affect the quality of life for citizens of the region and are not likely to improve without the commitment of both local residents and policymakers.

## Housing Problems of Special Populations

As in other parts of the nation, elderly, female-headed, and minority households living in the small towns and open country of the Midwest are most at risk of inadequate housing and housing affordability problems. Table 5.2 summarizes comparisons of the demographic characteristics and housing situations of these groups in the nonmetropolitan Midwest relative to the United States overall.[6]

Whether in the rural Midwest or in the United States as a whole, a large proportion of these special populations is poor. The mean monthly incomes of midwestern households were slightly lower than those of U.S. households as a whole, and the poverty rates were somewhat higher. For example, in the Midwest, 42% of all households headed by women and nearly one in three minority households (32%) fell below the poverty level, compared with nationwide proportions of 40% and 28%, respectively.

Table 5.2 also confirms that housing inadequacy generally was not a major problem for elderly, female-headed, and minority households in the rural Midwest and, in fact, was lower in the region for these populations than in the United States overall. (However, it should be noted that 1 in 10 rural minority households in the region was living in housing categorized by HUD as either moderately or severely inadequate.) The greater problem for these groups was the proportion paying more than 50% of household income on housing expenditures. Both female-headed and minority households in non-metro areas of the Midwest were slightly more likely than their U.S. coun-terparts to receive public assistance (44% and 29% vs. 40% and 21%, respectively). The proportion of minority households in the Midwest that received housing assistance was slightly greater than that of minorities living in the United States overall (25% vs. 16%).

The rural elderly were the most likely of the three population groups to be homeowners and to be satisfied with their housing and neighborhoods. Nearly 82% owned the homes they occupied, typically single-family de-tached dwellings. About 6% lived in mobile homes. Just 3% of the elderly's housing was characterized as inadequate, and 8% of the elderly received housing assistance. By contrast, just 42% of female-headed households were homeowners, 62% lived in single-family homes, and 13% resided in mobile homes. Only 6% of the homes were described as inadequate. More serious is the high percentage of those overpaying for housing. The average female-headed household paid $363 for housing. Nearly 17% of these households spent more than 50% of their income on housing.

Minority and female-headed households were the least likely popula-tions to own the homes they occupied. Less than half were homeowners. Compared with their U.S. counterparts, nonmetro minority households in the Midwest were more likely to be poor, to receive housing and public assis-tance, and to live in mobile homes.

**TABLE 5.2** Elderly, Female-Headed, and Minority Households: A Comparison of Selected Housing and Sociodemographic Characteristics in the Rural Midwest and U.S. Totals (percentages unless otherwise indicated)

|  | Nonmetropolitan Midwestern | | | U.S. Total Households | | |
|---|---|---|---|---|---|---|
|  | Elderly | Female Headed | Minority | Elderly | Female Headed | Minority |
| Age (mean years) | 75.2 | 37.6 | 43.0 | 74.6 | 39.6 | 44.2 |
| Children under 6 years of age present | 1.0 | 41.4 | 29.5 | 2.4 | 45.3 | 26.0 |
| Monthly income (mean dollars) | 1,800 | 1,561 | 2,282 | 2,073 | 1,881 | 2,581 |
| Poverty | 21.7 | 42.3 | 32.7 | 19.0 | 40.6 | 28.0 |
| Public assistance | 4.8 | 44.1 | 29.5 | 7.3 | 40.2 | 21.4 |
| Housing assistance (percentage) | 8.3 | 7.2 | 25.2 | 7.3 | 40.2 | 21.4 |
| Housing expense per month (mean dollars) | 284 | 363 | 356 | 381 | 508 | 584 |
| Housing costs more than 50% of income | 11.3 | 16.9 | 12.8 | 15.3 | 25.9 | 20.3 |
| Homeowner | 81.9 | 41.6 | 47.4 | 78.2 | 38.6 | 44.0 |
| Single-family dwelling | 80.9 | 61.6 | 64.0 | 72.2 | 54.5 | 55.3 |
| Mobile home | 5.5 | 13.2 | 13.2 | 6.3 | 6.4 | 3.4 |
| Inadequate housing | 3.4 | 6.2 | 10.4 | 6.0 | 10.7 | 11.7 |
| Housing satisfaction (mean) | 8.7 | 7.5 | 7.3 | 8.7 | 7.7 | 7.8 |
| Neighborhood satisfaction (mean) | 8.6 | 7.5 | 7.5 | 8.4 | 7.3 | 7.5 |
| Population estimate | 1,866,000 | 485,000 | 250,000 | 20,841,000 | 9,363,000 | 22,536,000 |

Two other special population segments in the rural Midwest with especially severe housing conditions, American Indians and migrant farmworkers, should be noted. (See Peck's chapter in this book [Chapter 7] for more detailed discussion of farmworker housing conditions.) The 1990 census reported 155,365 Native Americans living in the region. The number of Indians has shown rapid growth because of high birthrates and renewed interest in Indian culture, which has encouraged people to reclaim their heritage. Although the number of migrant farmworkers was not definitively counted, it was estimated that the migrant labor pool and dependents in the

12 midwestern states formed a population of 714,902 each year (Larson & Plascencia, 1993). Many Native Americans and migrant farmworkers have very low incomes, leading to problems of extreme overcrowding and very low housing quality. Both populations face persistent housing discrimination that further limits their residential choices. Moreover, migrant farmworkers have the additional problem of requiring decent and affordable accommodations in multiple locations during the harvest seasons.

## Vacation and Retirement Housing

In stark contrast to the housing problems experienced by some midwestern households are vacation housing and retirement community living. A popular midwesterner goal is to acquire a vacation home, often on a rural lake, that can become a retirement home. Approximately 8% of all midwestern housing units outside of metropolitan areas are seasonal units, although this number also includes some migrant worker housing. When people choose to retire in vacation areas, it can be a boon to small town economies. Over time, these vacation areas take on the characteristics of "naturally occurring retirement communities."[7]

The increased demand for services, however, can be a problem for the receiver community. Furthermore, in the northern Midwest, some seniors choose to be "snowbirds," traveling to southern or southwestern states in winter but returning to the Midwest for the summer. As such, retirees might not participate fully in community life in either location and might contribute relatively little to the local economy beyond their housing units.

## Environmental Quality Issues

Images of clean air and healthy environments in rural areas are only partly true. Most of the environmental issues and problems found in urban areas also can be found in the rural Midwest, sometimes with fewer effective control mechanisms in place. Inside of homes, poor air quality and presence of lead-based paint and radon often are associated with age and condition of housing units and their equipment. As indicated previously, the rural Midwest is second only to the Northeast in the age of its housing stock, and inadequate housing conditions constitute a significant problem.

Outside of homes may be found environmental quality conditions particularly associated with rural areas. Agricultural use of pesticides can be a hazard for people living near treatment areas. Although pesticide applicators must be trained in safe usage, mishaps do occur. The accidents can result in unintended spraying of areas, overuse to toxic levels, or spills that can contaminate soils and both surface water and groundwater. Agricultural runoff of pesticides, fertilizers, and animal wastes also can contribute to pollution of groundwater and surface water including private wells and aquifers used as drinking water sources. Nitrates from fertilizers and animal wastes are the major groundwater contamination concern, especially in areas of heavy fertilization in corn country. Volatile organic chemicals such as carbon tetrachloride (once used to spray grain) are a second concern. Other pesticides such as Atrazine also may occur as groundwater contaminants.[8] Children and older persons are most vulnerable to the effects of contaminated water.

In addition, rural areas of the Midwest are tempting locations for solid and hazardous waste disposal when political pressures prevent their location in metropolitan areas. Urban areas have shown a willingness to transport wastes long distances rather than find suitable disposal at home. Some rural communities have seen waste disposal as a viable new industry to replace others that are lost, not considering or appreciating the environmental problems this can cause for residents or the negative effect it can have on housing prices.

## The Changing Economy of the Rural Midwest

The economy in rural midwestern communities, once based on a mixture of agriculture, forestry, mining, and manufacturing, has changed and is changing rapidly. Communities once dominated by owner-operated farms have seen a long-term decline in farm numbers that was further jolted by the agricultural crises of the 1980s. As technology and farming practices have changed, agriculture no longer is the major employer in many counties. The number of farms has been falling for decades, and farm income now is less than 20% of labor and proprietor income in the vast majority of nonmetropolitan counties in the Midwest (U.S. General Accounting Office, 1993). Mining and manufacturing operations also have suffered from changing technologies and increased competition.

Besides growth in farm size made possible by mechanization, midwestern agriculture has felt the competitive pressures of the global economy. Many small farmers have been forced out of business or have had to supplement farm income with other employment. Some of their cropland has gone into other uses, but the majority has been absorbed by larger farming operations. At the same time, some farmers have increased income by raising specialty crops or by adding value through some on-farm processing. Nevertheless, year-round employment on farms has decreased. Fewer young people stay to take over their family farms, and retirees are not replaced.

When small farms are absorbed into large farms, local businesses that served them are hurt. The new, larger farms can negotiate for their supplies in larger centers. There are fewer farm operators shopping for their personal or family needs. Local businesses of all types see fewer farm dollars and can be forced out of business. As these businesses leave, the families they once supported move out and are not replaced. School populations decline, and local schools may be closed. Unless new enterprises replace those that are lost, community tax bases decline and government has fewer resources. This same pattern applies, although often in a more dramatically abrupt way, when mines or manufacturing plants close. Alternative job opportunities are few or nonexistent, and they often do not pay nearly as well. Disposable household income goes down, and main street businesses quickly feel the pinch. Whole communities sometimes die.

In addition, many of the Midwest's industries served agriculture as a source of supplies or of processing for land-based products. These industries were found both in cities and in small towns scattered throughout the region. Now, the number of older manufacturing industries has declined, while the number of recreation, tourism, and other service-based industries has grown. The existing housing stock often does not fit the new economic realities because it is in the wrong place or lacks the quality and/or amenities that buyers want.

Losses from economic downturns can have drastic consequences for the market prices of houses, which decline quickly. Just after a plant closing, buyers can find bargains, but there are fewer buyers than sellers. Sellers might not realize enough value to find comparable housing in other areas. They might feel that they cannot afford to sell or leave. Those who do stay might have fewer dollars for repairing and/or remodeling their homes, and housing quality might decline. Abandoned farm homesteads dot the midwestern countryside in mute testimony to the changes in agriculture.

In response to those changes, some rural communities have made efforts to diversify their economic bases by bringing in new development, using low costs and quality of life as inducements. Some have found that, after several lean years, their economies are stronger and healthier now than ever before because diversification has insulated them from bad times in one sector. Other communities have found it necessary to offer special tax or development concessions to draw or keep businesses. Industries planning to locate plants find communities and states competing for their businesses. They can keep wages low.

Availability of affordable housing for management and workers is an important development criterion, both for new industries and for expansion of existing industries. Without an adequate housing stock, businesses are reluctant to move to nonmetropolitan communities (Hartman, 1994). Some small towns have failed to realize the role of housing in maintaining and improving the vitality of their communities. Such communities, for example, might need to extend utilities to accommodate new housing. Without housing and necessary infrastructure, some employees might be forced to commute very long distances or might simply choose not to move with their companies. Because long commutes may translate into less free time for families and other obligations, failure to anticipate housing needs can be detrimental to community revitalization.

Tourism and recreation-related industries have been heavily promoted in the Midwest. Open spaces, clean air, and an abundance of lakes are enticements. Victorian houses offer charming settings for "bed and breakfast" operations. Indian gaming has exploded with success that has brought native and non-Indian employees to reservation areas where lack of housing has become a major concern. Nevertheless, many tourism and recreation jobs are near minimum wage, so housing affordability continues to be an important issue for employed persons.

## Conclusion

Life in rural communities throughout the Midwest has changed dramatically in the past two decades. Rural housing in the Heartland reflects the changing demographic characteristics and economic times in that region. Compared with other regions of the country, economic conditions for many are good, and new employment bases are being developed. However, as structural changes and competition have driven out of business small and middle-sized farms, mining operations, and manufacturing plants that supported local commerce, many small communities have declined and been left with surplus and deteriorating housing. The number and proportion of very poor and homeless households have increased. Particular types of households—the elderly, female-headed households, and minority families—are at high risk of housing problems.

To the extent that there are housing problems in the nonmetropolitan Midwest, they are driven primarily by three important characteristics: the aging of the population, the sparsity of people,[9] and the lack of jobs that pay well. In some counties in the Midwest, as many as one in three people is age 65 years or over, and the proportion of those age 85 years or over is inordinately high. The aging of the midwestern and farm region populations is said to foreshadow the problems that will face the nation over the next decade (American Association of Retired Persons, 1994). The fact that many of these older people and frail elderly live in remote rural places means that delivery of services to them will not be easily accommodated. Changes in rural economic conditions and agriculture have resulted in labor markets in which most jobs pay close to minimum wage rates and very few residents enjoy high pay.

It is likely that the rural landscape of the Midwest will continue to change. Future attention might need to focus on multicommunity strategies as small towns and rural places find it increasingly necessary to collaborate and make difficult decisions regarding resource allocation. Affordable and adequate housing alternatives must be found to ensure quality of life and the long-term economic vitality of these communities.

# Notes

1. For the purposes of this chapter, the Midwest includes the East and West North Central regions of the United States as defined by the U.S. Bureau of the Census (1990). It consists of 12 states: Illinois, Indiana, Iowa, Kansas, Michigan, Minnesota, Missouri, Nebraska, North Dakota, Ohio, South Dakota, and Wisconsin.

2. The data in Table 5.1 are from two sources: *Current Housing Reports* (HT150/95) and *American Housing Survey for the United States in 1995* (U.S. Bureau of the Census & HUD, 1997). The American Housing Survey is a longitudinal survey designed to provide detailed information on the same housing units and their current occupants from a national sample of about 50,000

interviews every other year. Interviewers return to the same housing units to study change in homes and households over time.

3. Nonmetropolitan households are those whose dwellings are not within a metropolitan statistical area (MSA) as defined by the U.S. Bureau of the Census (1990). MSAs are defined as the county or group of counties in which there is a city or twin cities of 50,000 residents or more and an urbanized area of 100,000 residents or more (U.S. Bureau of the Census, 1992b, p. A-8). Residents who live in small communities or in open country outside the city, but within the county, would be considered "metropolitan," whereas residents of a city of 40,000 not located within an MSA would be considered "nonmetropolitan."

4. Public assistance is cash or in-kind benefits received by the household from any means-tested program. Housing assistance is receipt of housing vouchers or certificates, public housing, other subsidized housing, interest subsidy, or energy assistance.

5. The number of mobile homes is increasing, especially in nonmetropolitan areas. However, mobile homes are not as common in the nonmetropolitan parts of the Midwest as in the South and West. Mobile homes make up 11% of the occupied housing units in all nonmetropolitan areas of the United States.

6. Table 5.2 employs raw data from the 1995 AHS downloaded from HUD's World Wide Web site. We define the elderly as households in which the respondents to the AHS are age 65 years or over. Female-headed households are women living with at least one child under 18 years of age and no spouse present. Minority households includes all those in which the respondents are non-white and/or Hispanic. Analyses are weighted to reflect the U.S. population. Whole weights were used rather than fractional weights.

7. See, for example, Hunt and Ross (1990).

8. This information was provided by the Region 7 Environmental Protection Agency office in Kansas City, Kansas.

9. The population densities of the regions of the United States range from 30.1 persons per square mile in the West, to 79.4 persons in the Midwest, to 98.2 persons in the South, to 313.1 persons in the Northeast (U.S. Bureau of the Census, 1990).

# Chapter 6

# The Hidden Homeless

*Mary Stover*

The condition of homelessness in the United States is widely viewed as an exclusively urban problem. The relative invisibility of rural homelessness is caused by several interrelated factors including public misconceptions about the causes and nature of homelessness, rural communities' lack of homeless service sites and visible congregations of homeless people, and lack of a national database on rural homelessness.

Federal policies typically define homelessness according to a person's or family's options rather than needs. For example, a woman who lives in a domestic violence shelter usually is classified as homeless. However, a woman who lives in an area where a domestic violence shelter is unavailable and flees domestic violence by doubling up with another household, no matter how precarious that household's situation becomes, typically is not considered homeless.[1] This predominant means of categorizing the homeless according to where they are sheltered affects homeless counts nationwide, but it especially affects rural areas.

If one views literal homelessness as one end of the housing spectrum and the condition of being adequately, affordably, and permanently housed at the other end, then persons who are near or at risk of literal homelessness,

AUTHOR'S NOTE: Christopher Holden of the Housing Assistance Council provided research and editorial assistance in the preparation of this chapter.

such as the woman who doubles up with another household to flee domestic violence, are somewhere in between. Individuals and families are at risk of homelessness if they are living in severely substandard or overcrowded housing or if they are experiencing severe housing cost burdens.

Literal homelessness often is episodic, whereas the condition of being without permanent adequate housing usually is longer term. The rural homeless typically move from one extremely substandard, overcrowded, and/or cost-burdened housing situation to another, often doubling or tripling up with friends or relatives. While housed in these precarious situations, the rural homeless do not meet the predominant interpretation of literal homelessness. They are, however, without permanent adequate homes.

## How Many Homeless Are There?

Accurately measuring the magnitude of homelessness is controversial. Measuring the magnitude of the problem in rural areas, in particular, has not been attempted on a national scale. Estimates of the size of the total national homeless population vary depending on the definition of homelessness and the methodology employed.

In its *Priority: Home! The Federal Plan to Break the Cycle of Homelessness,* the Interagency Council on the Homeless (ICH, 1994, p. 21) cited an estimate of approximately 7 million people experiencing homelessness in the United States at some point between 1985 and 1990. Other homeless counts receiving wide attention have ranged from a low of 250,000 people to a least 3 million being homeless at any given point in time (Hombs, 1990).

Just as ICH's plan noted that "recent studies confirm that the number of persons [nationwide] who have experienced homelessness is very large and greater than previously known or acknowledged," recent state studies and reports from rural homeless assistance providers confirm that the problem also increasingly plagues smaller communities (ICH, 1994, p. 21).

## What Are the Causes of Rural Homelessness?

Although there is no single cause of homelessness, poverty is the "common denominator" among many interrelated risk factors. Unemployment, under-employment, and housing affordability problems are the primary factors that contribute to rural homelessness.

Observers of homelessness outside America's inner cities often assume that homelessness has spread to rural and suburban areas previously believed to be immune from homelessness. In fact, homelessness has not "spread" to rural areas. The population most likely to be homeless, the very poor, always have been disproportionately represented among communities outside the metropolitan areas. More than 9 million people, or 16.3% of all Americans living in nonmetropolitan areas, had incomes below the poverty level in 1990.

After poverty, a lack of affordable housing is the most often cited cause of homelessness. In 1989, one in five rural households paid more than 30% of its income for housing, leaving little left over each month to pay for other necessities such as food and medical care (Housing Assistance Council [HAC], 1994b). A given individual's or family's likelihood of being homeless is increased if drug or alcohol abuse, chronic health problems (including mental health problems), and/or disabilities are present. Domestic violence, alone or compounding other problems, is a frequently cited cause of homelessness among women and children. A compilation of local surveys conducted nationally between 1981 and 1988 showed that in those years, 24% of the visible adult homeless population (primarily urban shelter residents) had spent time in a mental hospital, 27% were addicted to alcohol at the time of the survey, and 29% had spent time in residential treatment for substance abuse (Jencks, 1994, p. 22). Although it is apparent that substance abuse and mental illness are factors in rural homelessness, the data on these problems in the rural homeless population are sparse, and nonmetro/metro differences are inconsistent among statewide studies of homelessness. Statewide studies indicate that family conflict plays a more visible role in rural homelessness than in urban homelessness. A 1994 study of homelessness in Kentucky found that 48% of rural homeless women in the state, who made up the majority of Kentucky's rural homeless adults, were victims of domestic violence (Adams, 1994, p. 4). Rural assistance providers frequently note that lack of transportation increases vulnerability to homelessness, especially because remote areas rarely have public transportation connecting residents to jobs and services.

## Who Are the Rural Homeless?

State studies indicate some common characteristics of rural homeless populations when compared with urban homeless populations. The rural homeless population tends to be younger and includes fewer men, more women and families, and fewer minorities than does the urban homeless population (HAC, 1991, p. 23). The smaller number of minorities in the rural homeless population is not surprising given that it mirrors the geographic population distribution as a whole. Rural homelessness, however, disproportionately affects minority populations. Migrant workers and American Indians also are more likely to be among the homeless in rural and nonmetro areas (HAC, 1991, p. v). This reflects the higher poverty rates found among minority groups in rural and nonmetro areas in comparison to those found in urban areas.

A study of homelessness in Grand Island, Nebraska, illustrated some of the aforementioned observations. Grand Island is a nonmetro city of approximately 40,000 residents located in a predominantly agricultural region. In 1994, a total of 132 homeless individuals were counted in the city. A total of 51 of those counted as homeless were children. Whites made up 75% of the homeless counted in Grand Island, with the remaining 25% of Hispanic or other ethnic origin (Payne, 1994, pp. 1-5). Children were predominantly accompanied by single parents. A local social service provider noted that many of the homeless women with children were victims of domestic violence.[2]

## What Are the Solutions to Rural Homelessness?

As is true in urban areas, the ultimate solutions need to address not only individuals' shelter needs but also the underlying causes of homelessness, most notably poverty and lack of affordable adequate housing. In addition, homeless assistance providers must combat a wide array of ancillary problems exacerbated by homelessness such as deteriorating medical conditions, substance abuse, and domestic violence.

Homeless service providers in small towns and rural communities face a number of interrelated obstacles to establishing shelter and services. First, because the rural homeless often are hidden, it is hard to convince community members that a problem exists. When the homeless are identified, they often are perceived to be "outsiders" or people who are new to the community and brought their problems with them. Second, dis-economies of scale make start-up and operational costs for the many services needed difficult to justify because they serve relatively fewer people. Limited federal, state, and charitable funding usually is targeted to areas with the largest visible populations of homeless people. Thus, the bulk of efforts to assist the homeless are developed in urban areas.

The Stuart B. McKinney Homeless Assistance Act of 1987 created programs providing food, emergency shelter, transitional and long-term housing, health and mental health services, job training, education, and other support services to address the needs of homeless people. Although the portion of federal funding that makes it to rural communities varies from program to program and year to year, rural homeless assistance providers have not generally received their "fair share" of federal funding. Between 1987 and 1991 when funding was distributed on a solely competitive basis, nonmetro areas received only 5% of supportive housing for the homeless funds and 3% of Section 8 single-room occupancy funds. Only 2% of health care for the homeless funds and 3% of job training for the homeless funds reached nonmetro areas (HAC, 1995). Since 1994, the U.S. Department of Housing and Urban Development (HUD) has implemented a "continuum of care" planning process whereby funding for HUD homeless assistance programs is made available to states according to an allocation formula. While it is apparent that the continuum of care process has increased rural communities' access to federal funds, it is also apparent that competition for scarce resources remains, and rural homeless assistance providers continue to receive funds that are disproportionate to need.

A third difficulty faced by rural homeless assistance providers concerns the range of services necessary to serve the rural homeless. The needs of homeless individuals and families in small towns and rural communities are just as diverse as those in urban areas. Shelter, medical care, job training, counseling, and child care services frequently are required by homeless individuals and families. Because rural populations are smaller and geographically dispersed, providing a wide range of services becomes more challenging and expensive.

A fourth and related problem is that many nonmetro and rural service providers lack the program capacity to serve the homeless in their regions adequately. Community action agencies, churches, and small nonprofit organizations are the primary providers of homeless assistance in rural areas. Their limited capabilities and resources restrict the range of services available. Small communities often lack hospitals, public housing, public transportation, and public assistance offices that are taken for granted in urban areas. Local governments generally are inadequately staffed and ill equipped to provide homeless services.

When resources are available, however, the obstacles to service delivery in rural areas encourage the design of innovative delivery systems. Because the number of homeless people in a given community is small and congregate shelter usually is viewed as inappropriate, providers in rural areas have more incentive to focus on homeless prevention and permanent "re-housing" options. Individual providers' limited capacities encourage networking with other providers. Coordination and cooperation among service agencies lessens the obstacles posed by diseconomies of scale, diversity of assistance needs, and limited program capacities. Thus, necessity has resulted in many smaller communities providing models of service delivery cooperation to communities of all sizes.

States play an important role in the solution to rural homelessness. State resources should be directed toward rural areas that are unable to get funds directly from federal sources or generate their own funds locally. As cuts are made to federal housing, and as homeless assistance programs and welfare reforms drastically limit economic assistance, states' involvement in meeting rural homeless assistance needs will become even more crucial.

## Conclusion

According to *Priority: Home!,* the mention of homelessness in rural areas has been "all but absent in academic and policy debates" (ICH, 1994, p. 50). Rural homeless families and individuals are in no less need of housing and services than are those who live in urban areas where shelter for the homeless is more readily available and visible. Many literally homeless people are hidden. They do not congregate at city parks or shelters. They are in campsites in the woods, in abandoned barns and buses, and even in caves and chicken coops. They cannot be counted in soup kitchen lines if there is no soup kitchen nearby to serve them or if they do not have the transportation necessary to get there. We hold in America a bucolic image of life in rural communities, but this image is partially maintained by the continued invisibility of our rural citizens who are most in need.

# **Notes**

1. A homeless person is defined in the Stuart B. McKinney Homeless Assistance Act of 1987 as a person who

1. lacks a fixed, regular, and adequate nighttime residence or

2. lives in

(a) a shelter,

(b) an institution other than a prison, or

(c) a place not designed for or ordinarily used as a sleeping accommodation for human beings. (Hombs, 1990, p. 69)

2. Coordinator, Grand Island Community Help Center, HAC research telephone interview, August 3, 1995.

# Chapter 7

# Many Harvests of Shame:
# Housing for Farmworkers

*Susan Peck*

Seasonal and migrant farmworkers who plant, tend, and harvest the nation's agricultural bounty are mostly poor, minority, foreign born, and undereducated. They also are mobile, following the crops, sometimes alone and sometimes with spouses and children. Few agricultural areas of the United States provide year-round employment for farmworkers. Most work is seasonal, with some crops requiring fieldworkers for as little as several weeks, others requiring them for much longer periods, and still others requiring temporary workers at different times of the year. These seasonal workloads lead to a migratory labor force that follows the crops across state and national borders.

This flow of workers has no parallel in other industries, providing farmworkers with both a unique identity and a unique burden. Laden with household belongings and, often, family members, farmworkers are our modern-day nomads, in constant search of work and shelter. Yet, the communities and employers that seek farmworkers' aid in planting or harvesting might deny these same workers decent wages and good living conditions. One important concern for farmworkers is their need for decent and affordable housing. Too often, farmworkers live in shacks, barns, or chicken coops;

along riverbeds; in hand-dug caves; or amid unsafe electrical wiring, raw sewage, and polluted water.

Farmworkers have few resources to secure decent housing, and many growers are getting out of housing, finding it too costly and regulated. Rural communities often ignore these temporary workers, leaving to nonprofit sponsors and government agencies the task of obtaining funds and developing adequate housing. These groups look primarily to the federal government for help, and some might benefit from a handful of state programs, but the problem is large and the resources are small. A much greater commitment on the part of government is essential.

## Farmworkers in the United States

Farmworkers in the United States are increasingly Latin American immigrants. Many are undocumented and reliant on agriculture for most of their low incomes, according to national surveys required by the Immigration Reform and Control Act of 1986 (IRCA; U.S. Department of Labor, 1993). The National Agricultural Workers Survey (NAWS), although limited to workers performing farm labor on the day surveyed, provides the most comprehensive data now available on the nation's farmworkers.[1] The NAWS reveals a population that is poor and mostly unskilled, with limited education and English language ability, and dependent on agriculture for most of its income.

Interviews with farmworker housing organizations nationwide confirm this picture. Increasingly, the farmworker population of the United States is largely Hispanic. On the "Delmarva Peninsula" of Delaware, Maryland, and Virginia, the farmworker population more than a decade ago was mostly African American and Caribbean. By 1992, the population was nearly entirely Hispanic (84%), with a growing number of Central Americans, often undocumented. Most of the Mexican and Mexican American workers travel with their families, whereas the black population is mostly single males. Many Haitian workers, once prominent in the migrant stream, are "settling out" in local communities.

In San Diego County, California, entire families from Oaxaca, Mexico, now live in mountain encampments. Many are undocumented and have limited job and English language skills. Along the Mexican border in Texas and New Mexico, workers cross the border daily to work in the fields. Whole families with young children leave their home base just north of the border to travel the migrant stream to the Northwest, to the upper Midwest, and through the Mid-Atlantic states to the Northeast. More and more, according to nonprofit sponsors interviewed, these workers are traveling with families and are non-English speaking. Many cannot read in any language.

In some places, such as the agricultural areas of Oregon where the growing season can be 9 or more months of work, communities are becoming more Hispanic as farmworker families decide to stay. Where there are severe housing shortages and the growing season is short, such as in Colorado, male migrants often leave their families at home in Mexico or Texas.

In California, the nation's leading farm state, crop production is concentrated in high-value vegetables, fruits, and horticulture, requiring significant amounts of hand labor.[2] Estimates suggest that about 700,000 hired workers perform farm labor in California's agricultural sector at some point in the course of a single year (Gabbard, Kissam, & Martin, 1993). The NAWS survey found that about 92% of current California farmworkers were foreign born, a sharp increase from the 50% estimate of 20 years ago. Only about 9% of California farmworkers were not authorized to work in the United States. California agriculture now depends more on foreign-born farmworkers than at any time in this century. Moreover, the U.S.-born children of immigrant farmworkers do not want to do hired farmwork. Therefore, this labor force will replicate outside of the United States. The NAWS also disclosed that most farmworkers were young Mexican men who first came to California sometime after 1978 but that nearly half migrated each year from their usual places of residence to secure farm employment. As of 1994, the migrant proportion among the California farmworker population (47%) was higher than that found in 1983 (39%) (U.S. Department of Labor, 1994, p. 15). On the other hand, fully half of California fieldworkers are year-round U.S. residents. Newly arriving migrants are, increasingly, indigenous peoples from southern Mexico and Central America. Runsten and Kearney (1994) found 50,000 Mixtec Indians from the Mexican state of Oaxaca in California agriculture. Many speak only the Mixteco dialect and do not speak or read Spanish. This increased diversity presents new challenges to service providers.

In addition, farmworkers have extremely low incomes. According to the U.S. Department of Labor (1993) report on the NAWS, the annual median personal income of California fieldworkers was between $5,000 and $7,500. At least 48% lived in poverty, mainly caused by low wages and temporary employment. Yet, most of the state's farmworkers tried to combine a series of short-term farm jobs. Just 10% of fieldwork was performed by persons in the labor force on a seasonal basis. Despite their incomes, only 13% of farmworkers received any type of needs-based government services. The most frequently used service was food stamps (11%). Just 2% received Aid to Families with Dependent Children, and 3% reported public (low-income) housing assistance.

## Housing Needs—and Solutions—for Farmworkers

There are several critical issues in farmworker housing—the types of housing needed (single or multifamily, barrack-style, rental, or homeownership), the ownership entity (grower, nonprofit sponsor, or farmworker association), housing life span (seasonal or year-round), the level of subsidy needed for affordability, and social service needs (health, education, translation, job training, etc.). Developers must decide issues such as whether to build in the farmworkers' "home base" or where the workers live year-round, even if the labor is only seasonal. If the farm labor population is mostly migratory, then developers must know whether the workers are traveling alone or with family members, the length of time units will be occupied, and whether "winterization" or some lesser standard is needed.

The special circumstances and growing diversity of the farmworker population present substantial challenges to both affordable housing developers and service providers. Farmworker housing developers interviewed for this chapter raised the following key development issues: (a) land, infrastructure, and environment; (b) community response; (c) developer capacity; and (d) development costs versus resources. These issues are like those that would be raised by other low-income housing developers. However, the makeup and lifestyle of the population to be served—farmworkers—give these development issues a unique identity.

### Land, Infrastructure, and Environment

Farmworker housing development in many states is part of debates over loss of agricultural land to housing and building in areas of environmental hazards. If sites with pesticide contamination or fuel storage tanks are the only available or affordable building areas, then developers must factor in the potentially substantial costs of mitigating hazards. Housing lenders generally do not want to pay for mitigation. Unless grant funds can be obtained, the cleanup costs might render many developments financially unfeasible for the eventual occupants.

Although funding agencies such as the Rural Housing Service (RHS) (formerly the Farmers Home Administration [FmHA], part of the U.S. Department of Agriculture [USDA]) prefer that housing for farmworkers be situated near essential services, many rural communities lack adequate water and sewer infrastructure to support new development. The alternative is septic tanks, but finding appropriate soils on adequately sized drain fields that do not consume too much agricultural land is difficult at best.

RHS farm labor housing (Sections 514 and 516 loans and grants) is that agency's only program that may be used in a nonrural setting. This would seem to favor urbanizing agricultural areas such as San Diego County, where some 10,000 or more people, many of whom do farmwork, are known to live outdoors in the hills surrounding the city. However, urbanization also brings differing standards for development from what might be acceptable in a more rural setting. Multifamily zoning anywhere in the county is difficult to obtain. This traditional land use issue is closely aligned with majority community biases, providing another impediment to the development of housing for farmworkers in areas near their work.

## Community Response

Local community response to farmworker housing includes much "NIMBYism"—the "*not in my back yard*" sentiment. Many communities encourage an influx of agricultural workers during harvests but will oppose efforts to improve the workers' housing, particularly if improvements will encourage year-round residency. Red flags are raised—overcrowded schools (even though children of agricultural workers might already attend the local schools), criminal activity, and undocumented workers. Some developers attempt to counter this opposition by carefully selecting sites that do not require any public hearings on easements, conditional use permits, or variances. This is difficult in many rural and urban areas where communities are attempting to raise revenues through the land use and permit process.

Opposition to farmworker housing improvement is widespread but not universal. Colorado growers have formed partnerships with local and state government to finance and manage housing that will encourage reliable workers to return to their areas each year. This support, according to Colorado Housing Inc., occurs in areas that are mostly agricultural. Where the economy is more diverse, NIMBYism is more prevalent. Tierra del Sol (TDS) in New Mexico uses an educational process, emphasizing to the community and press the welfare of farmworkers' children. A local, single-purpose housing sponsor is created for each TDS project, with community leaders serving on the sponsor's board of directors. Local officials now call TDS with requests to take over management of poorly run farm labor camps.

## Developer Capacity

Farmers once provided much of the housing occupied by farmworkers, but with increased enforcement of health, safety, and housing laws, grower-provided housing has declined. In California, the number of state-licensed labor camps decreased from 5,000 in 1968 to 1,000 in 1994, coinciding with increased penalties for violations of that state's Employee Housing Act. Implementation of IRCA also coincided with a shift from direct hiring to agreements with labor contractors to provide workers to the growers—and to meet their housing needs (Harrison, 1994). Although individual growers or grower associations are eligible to obtain RHS loans for farm labor housing, the trend has been toward non-employer-provided units. Public housing authorities, nonprofit housing developers, and some farmworker associations and migrant health clinics have emerged as sponsors of housing for farmworkers.

Some years ago, FmHA (now RHS) began funding regional technical assistance providers to encourage the formation of local farm labor housing sponsors in needy areas. These providers say that it is hard to find willing sponsors. The process of preparing one or more funding applications, winning approvals for local improvements/services and federal funding, and finding suitable land is time-consuming. Very few local organizations have the money to support staff during the several years it might take to get a development approved, built, and occupied. Technical assistance is welcome, but the local sponsor needs to sustain itself, and "core operating money" is the most difficult to raise. If a project reaches the approval stage, then an

RHS fee might be available for the upfront costs of application, but the fee might not be sufficient.

### Development Costs Versus Resources

In 1980, a federally funded migrant housing needs assessment concluded that more than 750,000 units of housing were required to meet the unmet national need (Inter-America Research Associates, 1980). However, funding continues to decline or remain flat, and no new federal resources have been created. Current resources include (a) technical assistance funding from the Department of Labor, which enables longtime grantees of the department to assist others in developing farmworker housing; (b) Department of Health and Human Services funds through the Community Services Block Grants, available in a national competition for a small number of farmworker housing developments; and (c) farm labor housing loan and grant programs of RHS.

The Sections 514 and 516 loan and grant program for development of year-round or migrant housing is the nation's primary response to farmworker housing need. The program provides up to a 90% grant to nonprofit and public sponsors of farmworker housing. (The remaining 10% may be loaned at 1% interest; growers or their associations are only eligible for loans.) The program also helps with rental assistance for most units. Fiscal year 1997 funding of $25 million for the program assisted the development of only about 334 units nationwide, compared with 2,500 units assisted in fiscal year 1979 with funding of $68.5 million. A recent survey by the Housing Assistance Council (1997c) found pending or soon-to-be-filed applications for more than $134.5 million in 514/516 funding in 1998. However, only $25 million was available in the program for 1998. In the absence of improved funding at the federal level, several states have created programs specifically to aid in the development or rehabilitation of farmworker housing or have made general housing programs available for this purpose.[3]

## Conclusions

Farmworkers are one of the most economically disadvantaged groups in U.S. society. Inadequate housing clearly is a major problem faced by farmworkers, yet very little beyond compelling anecdotal evidence is really known about these housing conditions. Several things are needed:

- ➡ A thorough analysis and study of the farmworker population's housing conditions
- ➡ An evaluation of the experience of RHS's farm labor housing program
- ➡ Continuation of the USDA/RHS housing program at least at current levels, with possible expansion

## Notes

1. For a discussion of farmworker data sources, particularly those that provide some information on housing conditions, see Housing Assistance Council (1996).

2. Information presented in this section is drawn from Villarejo and Runsten (1993).

3. For a discussion of state efforts, see Housing Assistance Council (1986b, 1992).

# Chapter 8

# Housing for the Rural Elderly

*Joseph N. Belden*

The elderly occupy a special place in rural and nonmetropolitan America.[1] They are the people who remain behind in the small towns and countryside when the children and grandchildren travel or move to the city in pursuit of better opportunities. A popular view of the elderly is that they are relatively well off, with generous social security and Medicare benefits financed by younger workers. But in nonmetro areas, economic need among the elderly is actually higher than that among the nonelderly. Housing also is a vitally important need for older people in both urban and rural areas. Housing for senior citizens in the countryside has certain characteristics that differ from city and suburban conditions in terms of both needs and resources. The clear challenge is to be able to offer a continuum of housing choices for the elderly, particularly those with low incomes. Many cities and suburbs have successfully provided this range of available choices—a continuum proceeding from repair and rehabilitation for homeowners, to rental housing, to assisted living and long-term care. But rural and nonmetro places frequently have been less successful because of a lack of resources and capacity. This chapter examines the housing situation of rural and nonmetropolitan elderly, describes some federal programs designed to preserve or provide housing for low-income seniors, and offers suggested solutions.

Because of data limitations, this chapter analyzes population and poverty figures for nonmetropolitan areas but not rural areas. Housing data are available (and reported here) for both rural and nonmetropolitan places.

## Characteristics of the Nonmetropolitan Elderly

In 1995, 31.5 million people in the United States were age 65 years or over. This was 11.9% of a total population of 263.7 million. But in nonmetro areas, 13.9% of the population was age 65 years or over. A total of 7.2 million elderly people lived in nonmetro places. People age 75 years or over also made up a disproportionate number of the nonmetro population. In the suburban counties of metropolitan areas, persons age 75 years or over comprised 4.7% of the population. The same age group made up 5.9% of the population in nonmetropolitan areas. Table 8.1 shows these data. African American and Hispanic seniors in nonmetro places had very high poverty rates—46% and 33%, respectively. These were much higher poverty rates than those for the same groups in central cities and suburbs.

The nonmetro elderly also tend to be more likely than their counterparts in central cities and suburbs to suffer poverty conditions. As Table 8.2 shows, in 1995, there was a higher proportion of very poor elderly people (with incomes below 50% of the poverty threshold) in central cities. However, the elderly with incomes at or below 100% or 200% of the poverty threshold were more likely to live in nonmetropolitan areas. This was particularly true for people age 75 years or over. For this "older old" population in nonmetro places, more than 57% were below 200% of the poverty line. This compares with 42% in the suburbs and 49% in the central cities. The poverty threshold for elderly households in 1995 was $7,309 for one person and $9,219 for two people (Baugher & Lamison-White, 1996). This was, of course, a very low income for the mid-1990s. Clearly, a large number of the elderly live at or near poverty and have very limited resources.

A popular perception about the elderly is that they are better off than the general population. In metro areas, the poverty rate in 1995 was 13.4% for the total population and 11.7% for persons age 75 years or over. But in nonmetro areas, these proportions were reversed; the poverty rate was 15.6% for the total population and 17.5% for persons age 75 years or over.

**TABLE 8.1**  Elderly Persons by Location in United States, 1995 (persons in thousands)

|  | *Metro/Central Cities* | *Metro/ Suburbs* | *Nonmetro Areas* |
|---|---|---|---|
| All persons | 79,002 | 133,012 | 51,718 |
| Persons age 65 years or over | 9,218 | 15,243 | 7,197 |
| Percentage | 11.7 | 11.5 | 13.9 |
| Persons age 75 years or over | 4,064 | 6,280 | 3,044 |
| Percentage | 5.1 | 4.7 | 5.9 |

SOURCE: Baugher and Lamison-White (1996).

**TABLE 8.2**  Poverty Status of Elderly Persons by Location in United States, 1995 (persons in thousands)

|  | *Metro/Central Cities* | *Metro/ Suburbs* | *Nonmetro Areas* |
|---|---|---|---|
| Persons age 65 years or over | 9,218 | 15,243 | 7,197 |
| Percentage under 50% of poverty level | 2.5 | 1.7 | 1.7 |
| Percentage under 100% of poverty level | 12.4 | 8.1 | 13.1 |
| Percentage under 200% of poverty level | 42.1 | 34.1 | 48.1 |
| Persons age 75 years or over | 4,064 | 6,280 | 3,044 |
| Percentage under 50% of poverty level | 2.9 | 2.2 | 2.4 |
| Percentage under 100% of poverty level | 14.4 | 10.0 | 17.5 |
| Percentage under 200% of poverty level | 49.3 | 42.2 | 57.2 |

SOURCE: Baugher and Lamison-White (1996).

## Housing Conditions

The elderly in nonmetro and rural areas in 1995 were more likely to be homeowners, to live in mobile homes, to have physical problems with their units, and to have fewer resources to meet low-income housing needs.

Rural areas in 1995 had 6.0 million occupied housing units with elderly householders, compared with 5.8 million in nonmetro areas. These totals were up from 5.8 million and 5.6 million, respectively, in 1993. Tables 8.3 and 8.4 compare those two years, using data from the American Housing

**TABLE 8.3**  Occupied Housing Units With Elderly Householders in United States, 1993 (in thousands)

|  | Total | Owner Occupied | Percentage Owner Occupied | Renter Occupied | Mobile Homes | Percentage in Mobile Homes |
|---|---|---|---|---|---|---|
| Total United States | 20,438 | 15,767 | 77.1 | 4,671 | 1,288 | 6.2 |
| Metropolitan areas |  |  |  |  |  |  |
| Central cities | 6,039 | 3,969 | 65.7 | 2,070 | 92 | 1.5 |
| Suburbs | 8,646 | 7,185 | 81.2 | 1,661 | 606 | 7.0 |
| Nonmetropolitan areas | 5,553 | 4,613 | 83.1 | 940 | 569 | 10.2 |
| Urban areas |  |  |  |  |  |  |
| Total | 14,709 | 10,741 | 73.0 | 3,968 | 493 | 3.4 |
| Nonmetropolitan | 2,182 | 1,681 | 77.0 | 501 | 106 | 4.9 |
| Rural areas |  |  |  |  |  |  |
| Total | 5,729 | 5,026 | 87.7 | 703 | 774 | 13.5 |
| Suburbs | 2,334 | 2,076 | 88.9 | 258 | 311 | 13.3 |
| Nonmetropolitan | 3,371 | 2,932 | 87.0 | 440 | 463 | 13.7 |

SOURCE: U.S. Bureau of the Census and U.S. Department of Housing and Urban Development (1995).

Survey. Most of the elderly-occupied units were in the South and Midwest. In rural areas, 88% of these units were homeowner occupied, compared with 83% in nonmetro places. Some other important characteristics of the non-metro and rural elderly's housing in 1995 included the following:

➤ Fully 798,000 rural and 579,000 nonmetro units were mobile homes.

➤ About 150,000 rural and 140,000 nonmetro homes lacked some or all plumbing facilities.

➤ More than 1 million units in both rural and nonmetro places did not have adequate heating equipment.

➤ Only about 3.4 million rural and 4.2 million nonmetro units were part of public water systems.

➤ Fully 485,000 rural and 468,000 nonmetro units had moderate or severe physical problems with plumbing, heating, electrical systems, upkeep, hallways, and/or kitchens.

➤ Fully 1,002,000 rural and 872,000 nonmetro units had exterior water leakage in the previous year.

➤ Fully 119,000 rural and 149,000 nonmetro units had signs of rats.

**TABLE 8.4** Occupied Housing Units With Elderly Householders in United States, 1995 (in thousands)

| | Total | Owner Occupied | Percentage Owner Occupied | Renter Occupied | Mobile Homes | Percentage in Mobile Homes |
|---|---|---|---|---|---|---|
| Total United States | 20,841 | 16,299 | 78.2 | 4,542 | 1,320 | 6.3 |
| Metropolitan areas | | | | | | |
|   Central cities | 5,916 | 3,964 | 67.0 | 1,952 | 93 | 1.6 |
|   Suburbs | 9,168 | 7,491 | 81.7 | 1,677 | 648 | 7.1 |
| Nonmetropolitan areas | 5,756 | 4,843 | 84.1 | 913 | 579 | 10.1 |
| Urban areas | | | | | | |
|   Total | 14,865 | 10,990 | 73.9 | 3,875 | 522 | 3.5 |
|   Nonmetropolitan | 2,229 | 1,724 | 77.3 | 505 | 101 | 4.5 |
| Rural areas | | | | | | |
|   Total | 5,975 | 5,308 | 88.8 | 667 | 798 | 13.4 |
|   Suburbs | 2,419 | 2,163 | 89.4 | 256 | 320 | 13.2 |
|   Nonmetropolitan | 3,527 | 3,119 | 88.4 | 408 | 478 | 13.6 |

SOURCE: U.S. Bureau of the Census and U.S. Department of Housing and Urban Development (1997).

➡ Fully 228,000 rural and 209,000 nonmetro homes had toilet breakdowns in a 3-month period.

The nonmetro and rural elderly also have less access to rental housing. To some extent, this is a matter of preference. For all rural and nonmetro households, the homeownership rate is very high. But as people age, many want and need apartment living. Between 1993 and 1995, for the elderly in rural and nonmetro areas, the number of homeowner-occupied units and mobile homes went up, but the number of renter-occupied units declined.

## Solutions

In the face of the sort of need described here, an obvious solution is to provide housing assistance. However, the resources currently available are limited. In nonmetro areas, 1.287 million units were occupied by poor elderly households in 1995. Nearly the same number of units in central cities (1.309 million) were occupied by the elderly poor in the same year. However, there was a sharp disparity in the numbers of subsidized units. In central cities, 584,000 elderly households lived in public housing or other subsidized units. Less than half as many public or other subsidized units (257,000) were occupied by elderly households in nonmetropolitan areas. In addition, there were only 110,000 such units in rural areas (U.S. Bureau of the Census & U.S. Department of Housing and Urban Development [HUD], 1997). These wide disparities might show that the cities were better at grant seeking and that there was more opposition to low-income housing in small towns.

The federal programs used for seniors in these communities included repair and rehabilitation, HUD Section 202 housing for the elderly, public housing owned by local housing authorities, and privately owned housing with rental and construction subsidies. For example, the Rural Housing Service (RHS) Section 504 home repair program provides loans of up to $15,000 and grants of up to $5,000 to individuals. The grants are available only to the very low-income elderly to remove health and safety hazards. Grants and loans often are combined, with loans available for up to 20 years at 1% interest. The Section 504 program has helped many very poor seniors get amenities such as indoor plumbing for the first time in their lives. For the past several years, funding for the combined loans and grants in Section 504 has been at about $60 million. In 1996, this amount repaired 11,400 units, mostly for the low-income elderly (HAC, 1997b).

The main federal housing construction effort today for the low-income elderly is the HUD Section 202 program. It provides capital grants to nonprofit sponsors for construction or rehabilitation of apartments and related facilities for persons age 62 years or over. A "set aside" of 25% of these funds goes to rural areas. Section 202 projects also may include financing for support services, and HUD Section 8 rental assistance may help support the rents of low-income tenants.

These programs have fared relatively well in the budget wars, but others have not. For example, the RHS Section 515 rural rental housing program builds apartments for low-income rural people. About 40% of this program finances multifamily units for the elderly, but there have been very steep reductions since 1994. In 1993, the Section 515 program financed 15,300 apartments. However, in 1996, only 1,900 were supported.

An aging baby boom population will, within one to three decades, cause the elderly population of the United States to explode. One implication of that change is that there clearly will be a growing need for more and better housing for the low-income rural elderly. Home repair to help seniors age in place and suitable apartments when needed are two ways in which to provide such housing. However, to provide those solutions, adequate public support is vital.

# Note

1. "Rural" and "nonmetropolitan" are not synonymous. The U.S. Bureau of the Census defines rural areas as either open country or places of less than 2,500 residents. Nonmetropolitan areas are those counties that lie outside metropolitan statistical areas. Metro areas consist of counties with central cities of at least 50,000 residents and surrounding contiguous counties that are metropolitan in character.

# Part III

# Public and
# Private Money

# Chapter 9

# The Role of the Federal Rural Housing Programs

*Art Collings*

Since 1950, the federal government has operated programs to help provide affordable housing in rural areas. Until 1995, the agency operating these programs was the Farmers Home Administration (FmHA), a part of the U.S. Department of Agriculture (USDA). In 1994, FmHA was eliminated as part of a USDA reorganization. FmHA programs were distributed among four agencies under an Under Secretary for Rural Development. Farmer programs were transferred to a new Farm Service Agency, and three rural development services were created. The former FmHA housing programs are now in the Rural Housing Service (RHS), one of the three Rural Development (RD) agencies.

The current RHS rural housing programs evolved[1] from the failure of U.S. Department of Housing and Urban Development (HUD) predecessor agencies and states to provide housing finance in small and remote rural areas.[2] Prior to 1961, FmHA's housing role was limited to farm housing, although the definition for a farm in the Housing Act of 1949 used a very liberal interpretation.[3] The original role or mission for FmHA housing, both in the 1949 act and as expanded in the Housing Act of 1961,[4] was to finance modest housing or housing repair for families that lacked their own resources or could not obtain other credit at affordable rates and terms. FmHA was to be the lender of last resort. The FmHA programs have been expanded and diversified but have essentially maintained the basic "no credit elsewhere" concept.[5] The programs essentially are carried out directly by RHS/RD staff through a network of state, district, and county offices.[6] Once in excess of 2,000, about 1,000 RD offices remain today, and that number is likely to decrease. These offices have provided local access to applicants and borrowers.

## Past and Current Programs

Since 1949, FmHA/RHS has financed or rehabilitated more than 2.7 million rural housing units. With more than $70 billion injected into rural economies, FmHA/RHS provides both housing and economic stimulus. For example, the National Association of Home Builders[7] estimates that Section 515 rural rental housing in fiscal year 1993 generated $326 million in wages within a $573 million program. The social role of this program cannot be discounted. As a direct result of the subsidies used, the Section 515 program can house the poorest of the poor. In 1995, nearly 88% of tenants in 380,575 units had incomes below 50% of the area median.[8] FmHA/RHS rural housing programs also often provide employment and help stabilize rural communities.

Two ancillary programs, Section 523 self-help grants and Section 521 rental assistance, help make the building programs more affordable. Section 523 covers administrative costs for housing constructed through mutual self-help. Under the self-help program, nonprofit and public organizations sponsor groups of families that help each other build homes. FmHA/RHS requires that 65% of work be done by the families, which results in significant construction savings. Smaller loans not only yield lower loan payments and/or increased affordability but also reduce the amount of interest subsidy needed, thus lowering cost to the government.

Section 521 rental assistance is a deep subsidy that helps tenants in Section 515 rental housing units or Sections 514 and 516 farm labor housing units keep their total shelter costs to 30% of adjusted income. Since its inception in 1978, this program has supported tenants in nearly 458,000 units.

Table 9.1 provides an overview of the various programs. Still another catalyst that enhances affordability in the Section 502 homeownership program is the deferred mortgage payment authority. It allows families with incomes below 80% of the area median to receive 33-year mortgage loans with as low as 1% interest. Families with incomes below 60% of the area

**TABLE 9.1**  FmHA/RHS Programs That Construct, Purchase, or Repair Rural
Housing Units From Inception Through Fiscal Year 1996

| Program[a] | Year of Inception | Dollars Obligated | Units |
|---|---|---|---|
| Section 502 | | | |
| Homeownership loans: Direct | 1950 | 50,304,172,721 | 1,926,841 |
| Homeownership loans: Unsubsidized, guaranteed | 1991[b] | 4,305,161,920 | 67,657 |
| Section 504 | | | |
| Very low-income repair loans | 1950 | 328,386,182 | 108,267 |
| Section 504 | | | |
| Very low-income repair grants | 1962 1977[c] | 323,611,035 | 97,415 |
| Section 514/516 | | | |
| Farm labor housing loans and grants | 1962[d] 1966[e] | 654,742,478 | 29,363[f] |
| Section 515 | | | |
| Rural rental housing | 1963 | 14,083,630,680 | 5,514,710[g] |
| Section 533 | | | |
| Housing preservation grants | 1986 | 220,699,580 | 48,142 |
| Section 538 | | | |
| Guaranteed multifamily housing | 1996 | 16,180,642 | 450 |
| Totals | | 70,236,585,238 | 2,793,205 |

SOURCES: FmHA reports (Report Code 205), Multi-Family Housing reports, and Automated Multi-Family Housing Accounting System reports compiled by the author.
NOTE: FmHA = Farmers Home Administration; RHS = Rural Housing Service.
a.  Section numbers refers to sections within Title V of the Housing Act of 1949, as amended.
b.  A total of 1,210 units produced during fiscal years 1977 to 1981.
c.  Section 504 grants were not funded from fiscal years 1966 through 1976.
d.  The first year of activity for Section 514 farm labor housing loans was 1962.
e.  The first year of activity for Section 516 farm labor housing grants was 1966.
f.  Davis-Bacon wage rates apply where Section 516 grants are used. This normally increases unit costs in most rural areas.
g.  This is from 26,205 loans. A total of 449,249 units in 17,857 projects remained in the Section 515 portfolio as of January 1995.

median that cannot afford the 33-year mortgages might be eligible for 38-year 1% loans. Families with incomes below 50% of the area median that cannot afford the 38-year loans might qualify by deferring payment on up to 25% of those loan payments. Unfortunately, the program was not funded after fiscal year 1995.

**TABLE 9.2** Combined Direct Section 502 and 515 Unit Production in Selected Years

| 1970 | 1975 | 1980 | 1985 | 1990 | 1994 | 1995 | 1996 |
|------|------|------|------|------|------|------|------|
| 68,028 | 115,942 | 115,347 | 69,032 | 41,032 | 38,874 | 18,204 | 17,796 |

With sufficient funds, RHS can play an indispensable role in providing housing for low-income rural families. Most RHS housing programs are targeted to families with incomes below 50% of the area median. These programs supplement, and do not compete with, other lending sources in rural areas. History has shown that many less targeted housing programs do not have the same impact in rural America. Poorly documented or authenticated claims indicate a lack of private sector mortgage activity in rural areas due to economies of scale, limited capacity, and/or lower incomes. Regardless, the result has been a crucial, if supplemental, role for the direct[9] FmHA/RHS programs.

The preamble to the Housing Act of 1949 states that every American family has the right to decent, safe, and sanitary housing in a suitable environment. The FmHA/RHS role is to help rural families achieve this goal. However, unit production is actually lessening as population growth and increasing poverty fuel the need for increased assistance. Combined Sections 502 and 515 unit production since 1970 illustrates the trend, as shown in Table 9.2.

Need remains, with nearly 1.5 million occupied substandard rural housing units and more than 2.2 million rural tenants paying more than 30% of their incomes for rent. But in dollar terms, support for subsidized low-income housing is declining. From 1994 to 1997, FmHA/RHS subsidized programs fell from $3.072 billion to $1.436 billion. At the same time, unsubsidized programs grew from $800 million to $2.3 billion.

FmHA/RHS traces its agency roots to the Resettlement Administration, the Farm Security Administration, and other federal relief initiatives of the 1930s. These efforts included many rural social programs such as camps for migrant farm workers, housing, new towns, and resettlement of large tracts into family-owned farms. After World War II, the Farm Security Administration came under attack as socialistic and evolved into FmHA in 1946. Its original mission of providing supplemental credit to family-size farmers expanded over time to include funding for water and sewer systems, community facilities, business and industrial loans, rural planning, industrial grants, and housing. An examination of FmHA's housing programs shows a gradual shift of the agency back to more socially oriented purposes. The 2.6 million units of housing financed since 1950 are a measure of that change.

The role of the agency has changed not only with authorizing legislation but also with the goals and policies of succeeding administrations. Currently, RHS tries to balance loan servicing with new originations. However, the agency does have credit history standards for Section 502 applicants that are more rigid than those of most banks. The interest credit subsidy in the Section 502 program has been replaced by a more shallow assistance scheme. Multifamily housing regulations permit borrowers to exclude tenant applicants on the basis of credit reports. The worthwhile aspects of the rural housing programs are best described in outline form in Table 9.3.

Targeting to those most in need has been a key factor in defining a rural housing role for FmHA/RHS. The agency's programs have evolved far beyond the original farm housing role, with changes mirrored in FmHA/RHS regulations as found in the Code of Federal Regulations.[10] They show a clear focus on those with limited incomes.

The FmHA/RHS rural housing programs have been successful, but their role is changing. The Clinton administration, as part of "reinventing government," strove for a smaller budget. Congress, facing a need for massive budget cuts, willingly accepted reduced housing spending for fiscal year 1995 and has been ready to slash deeper for the succeeding fiscal years. The shortage of direct Section 502 funding prompted RHS administrators to begin a program of shared loans with private lenders and state agencies. This leads to a blended interest rate that can force housing assistance to families with higher incomes.

## A Future Role

Recent Congresses have not yet decided on an RHS role. There has been discussion of converting the direct loans to guarantees. The Clinton administration has been moving cautiously in this direction because the cost of guaranteed Section 502 loans is 0.17% and that for direct loans is 14.18%.[11] Both branches of federal government are committed to centralized loan servicing. Another proposal would transfer rural and other housing programs to a quasi-public Federal Housing Corporation. The federal deficit and a

**TABLE 9.3** A Brief Description of Title V Programs of the Housing Act of 1949

| Section | Description | Term | Subsidy |
|---------|-------------|------|---------|
| 502 direct | Homeownership loans[a] | 33-38 years | Interest down to 1% |
| 502 direct | Homeownership loans[a] | 38 years | Interest down to 1%<br>Deferred mortgage payment |
| 502 guaranteed | Homeownership loans[b] | 30 years | Unsubsidized |
| 504 | Home repair loans[c] | 20 years | Flat 1% rate |
| 504 | Home repair grants[d] | — | Grants of up to $5,000 |
| 514 | Farm labor housing loans | 33 years | Flat 1% rate |
| 515 | Rural rental housing loans[e] | 50 years | Interest down to 1% |
| 516 | Farm labor housing grants | — | Grants of up to 90% of development costs |
| 521 | Rental assistance[f] | 5-year contracts | Subsidizes tenant shelter payments to 30% of income |
| 523 | Self-help grants | 2 years | Administrative cost for self-help sponsors |
| 525/509 | Capacity building grants | 1 year | Grants |
| 533 | Housing preservation grants[g] | 2 years | Grants for rehabilitation |
| 538 | Guaranteed multifamily rental housing loans | 40 years | 20% of projects receive limited interest subsidy |

a. Available only to those with incomes below 80% of the area median. A minimum of 40% of funds go to those with incomes below 50% of the area median.
b. Moderate income program (to 115% of the area median).
c. Limited to very low income (below 50% of the area median).
d. Limited to very low-income senior citizens.
e. Rental housing for low- and moderate-income tenants.
f. Deep subsidy for low- and very low-income tenants in Section 515 and farm labor housing.
g. Competitive grant program in which applicants develop how grant funds will be used—loans, grants, subsidy for housing rehabilitation, and the like.

mood toward less government have placed the rural housing programs, as well as the new RHS, at risk.

What is a proper role for RHS in rural housing? The author's recommendations are set out in what follows. They represent personal bias, tempered by longevity in the field.[12] The RHS role should include the following:

1. Serves as a direct lender of last resort[13]
2. Programs targeted to those with the lowest incomes living in deficient housing

3. A full array of resources to fit varying needs—homeownership and repair loans, grants for special needs, a mutual self-help program, cooperative loans, interest subsidy, rental assistance, mortgage deferral, and moratoria
4. Flexibility to address remote occupancy, migrant farmworker situations, and lending on Indian reservations
5. Uniform standards primarily to protect consumers but designed for simplicity
6. National policies and criteria adaptable to local situations but without weakening the right of consumers to a thoughtful national program
7. A field delivery system for single-family housing applicants, designed for accessibility to lower income, undereducated rural residents

RHS's direct lending role must be maintained. The families assisted in the direct loan program are not served by the private sector or by RHS's guaranteed loan authority. In fiscal year 1996, the average direct Section 502 loan borrower's income was 63% of the rural household median. The guaranteed Section 502 counterpart's income was 112.3% of the rural household median. It is the primary role of the agency to serve that lower end of the income spectrum. It is fair to note that consumer protections in the law, including the right to appeal adverse decisions and a payment moratorium in Section 502 as a result of circumstances beyond the borrower's control, are denied to those applying for or receiving guaranteed loans.

In short, the RHS role should be to implement the promise of the Housing Act of 1949—that all rural residents should be able to obtain decent, affordable housing in a suitable environment. To achieve this, there must be a continuing evolution of law and regulations and an increase, not a decrease, of funding.

## APPENDIX

## An Outline of Selected Provisions
## of Title V of the Housing Act of 1949

All sections noted are within this law. They represent social evolution of the rural housing programs and are listed in the order enacted.

*Section 502(a)(4)* permits refinancing an owners mortgage to avoid foreclosure.

*Section 502(e)* authorizes (mandates) escrowing of taxes and insurance.

*Section 502(a)(2)* authorizes 38-year loans for applicants unable to afford 1% 33-year terms and whose incomes are below 60% of the area median.

*Section 502(c)* restricts prepayment of Section 515 and Section 514 loans to prevent displacement of tenants.

*Section 502(d)* mandates that no less than 40% of Section 502 appropriations be available only to those with incomes below 50% of the area median.

*Section 502(g)* provides an authority to defer up to 25% of Section 502 payments for applicants within incomes at or below 50% of the area median.

*Section 504* authorizes loans and grants to very low-income homeowners for housing repairs and rehabilitation. There is a special emphasis on removing health and safety hazards. Since 1976, Congress, in appropriations language, has restricted grants to the elderly.

*Section 505(a)* provides for a moratorium on payments where the inability to do so was caused by reasons beyond the borrower's control.

*Section 509(d)* prevents alienation of land in the case of default of loans on tribal allotted or trust land.

*Section 509(f)* provides for a "set aside" of housing funds for 100 counties designated as underserved that also have a poverty level of at least 20% and an occupied substandard housing rate of at least 10%. Provisions are included for assistance in colonias. Section 509(f) also provides grants to develop capacity in these areas.

*Section 510(g)* establishes the right to appeal adverse decisions and requires the agency to provide regulations thereof.

*Section 514(a)(2)* establishes a flat 1% interest rate for farm labor housing loans.

*Section 515(e)(4)* permits 102% loans to nonprofit and public applicants to cover initial operating expenses.

*Section 515(p)* prioritizes occupancy for very low-income families when Section 521 rental assistance is available.

*Section 515(w)* authorizes a 9% "set aside" of allocated funds to nonprofit sponsors.

*Section 516(b)* provides farm labor housing grants of up to 90% of development costs.

*Section 521(a)(1)* provides interest subsidy down to 1% of Sections 502 and 515 loans.

*Section 521(a)(2)* authorizes rental assistance to subsidize the difference between 30% of tenants' income and rent including utilities.

*Section 532(a)* requires priority in processing applications for those with greatest housing need due to lowest incomes and residence in inadequate dwellings.

# Notes

1. Since passage of the Housing Act of 1949.

2. The Housing Act of 1961 expanded eligibility to nonfarm housing, using the census definition for rural areas—essentially open country and communities of up to 2,500 residents that are rural in character.

3. The Housing Act of 1949 defines a farm as producing or capable of producing not less than $400 worth of agricultural commodities at 1944 equivalents.

4. Public Law 87-70.

5. Public entities are exempt from "no credit elsewhere" requirements. In addition, "no credit elsewhere" is somewhat of a sham in the Section 515 rural rental housing program where that aspect is tied to financing necessary to produce rents affordable to FmHA eligible tenants.

6. Since 1992, there has been a growing emphasis on guaranteed housing loans. In 1995, the amount available for unsubsidized guaranteed Section 502 loans exceeded that for direct Section 502 loans.

7. Economic and Policy Division.

8. Data are from the FmHA Multiple Family Housing Occupancy Survey, January 1995. No subsequent data were available as of this writing. Only 380,575 of 422,289 occupied units reported income.

9. Programs authorized by Congress and directly administered by FmHA/RHS field staff.

10. 7 CFR, Parts 1806 to 1980 in the *Code of Federal Regulations.*

11. Budget Authority or the estimated cost of loans made in fiscal year 1997 over the lives of those loans (also expressed in percentage of the program appropriation). Data are from the appendix of the fiscal year 1997 budget of the U.S. government.

12. The author has worked with the FmHA/RHS rural housing programs since 1955, including 3.5 years developing FmHA housing policy for the agency, and has worked for the Housing Assistance Council for 21 years.

13. A lender of last resort provides credit assistance to those unable to obtain it from conventional sources or from their own resources.

# Chapter 10

# Privatizing Rural Rental Housing:
# The Prepayment Problem in RHS's
# Section 515 Program

*Robert J. Wiener*

From 1966 to 1987, in hundreds of small communities, loans made by the U.S. Department of Agriculture's (USDA) Rural Housing Service (RHS) to encourage private production of more than 400 low-cost apartment complexes were quietly paid in full before their maturity dates (U.S. General Accounting Office [GAO], 1988, p. 11). Prepayment of these loans, financed under Section 515 rural rental housing program, had the effect of completely "privatizing" the properties—releasing them from all federal government controls on admissions and rents. Units originally built to serve elderly and low-income tenants were converted to market rates with increased tenant rents and hardship. Before Congress placed a moratorium on prepayments at the end of 1986, more than 5,000 units were irretrievably lost to the program (GAO, 1988).

The emergence of the prepayment problem in RHS's Section 515 program can be located and explained within the context of a recurring struggle of competing ideological orientations regarding the appropriate direction and means of federal intervention in the housing sector. Some have defined this struggle in terms of conservative and liberal notions of government action (Lowi, 1979; Savitch, 1979). The majority of conservative political actors view the private sector as the most efficient and effective agent for producing and distributing housing, and resolving the shelter problems of lower-income families (Hays, 1985). Only those limited government actions that reinforce the direction the market system is taking on its own (e.g., low-interest loans, mortgage insurance, tax breaks) or stimulate noneffective demand are justifiable interventions. Interventions that restrict normal market transactions, or the normal laws of supply and demand, are seen as undue meddlings (Marcuse, 1986).

Liberals, on the other hand, advocate a more activist government role that restrains and channels the activities of the private market to achieve a fairer distribution of goods, services, and wealth. Within the housing sector, market inequality is best overcome through direct government financing, ownership, and control of housing. Public supports to encourage private investment and ownership by profit-motivated concerns are tolerated where the housing produced is deeply targeted to lower-income families for the longest possible terms. Government activism also is justified where it is necessary to rectify the unforeseen effects of government policies, programs, or administrative decisions that produce adverse market consequences and frustrate the public purpose of providing decent, affordable shelter. These are seen as necessary *meliorist* actions (Savitch, 1979) by an essentially benevolent state to minimize the social costs of poor housing, which the private market economy is unwilling to correct, or incapable of correcting alone (Marcuse, 1986).

This conservative/liberal dichotomy also informs the current debate over prepayment. Conservative proponents of unfettered prepayment argue that the great majority of owners, when faced with less economic conventional uses for their properties, will not prepay. Moreover, current project-based preservation strategies, besides being too costly, are too inflexible. A tenant-based solution predicated on the provision of portable rent subsidy vouchers (or certificates) when owners prepay will provide maximum choice to tenants "trapped" in housing that offers less desirable accommodations than the available alternatives. Buildings undergoing prepayment will retain their low-income character if they are able to compete for assisted tenants alongside market-rate properties. Existing tenants need not be concerned, as this inventory is "voucherized"—converted from a project-based system to a tenant-based system.

Liberals argue that the government must intervene in prepayment situations. If not an outright prohibition, conditions should be imposed that severely restrict prepayment and offer owners financial incentives to transfer their properties to tenant- and community-based, nonprofit owners who will keep the units affordable in perpetuity.[1] It is both necessary and appropriate for Congress to correct the unanticipated consequences of government decisions that contributed to the prepayment problem and violate the original intent that this housing remain affordable for at least the term of the financing. In the absence of such interventions, owners surely will opt to maximize their profits by prepaying and converting to more lucrative market-rate operations. Dismantlement of the project-based system in favor of tenant-based vouchers ultimately will reduce tenant choice, increase uncertainty, and result in the permanent loss of low-cost housing.

Today, in light of recent actions by Congress to restore owner prepayment rights and end project-based financial incentives to preserve the much larger inventory of U.S. Department of Housing and Urban Development (HUD) rental housing,[2] it is instructive to reflect on the little known, but important, lessons of the RHS prepayment experience. This chapter traces the legislative, regulatory, and programmatic developments that led to the RHS prepayment problem. It shows that the Section 515 program is firmly embedded within the liberal housing tradition and that prepayment is a natural consequence, in fact, the logical conclusion of the liberal strategies that emerged in the 1960s. The chapter concludes with thoughts on the failure of this approach to produce permanently affordable rental housing and the meaning of the RHS prepayment experience for HUD's older assisted housing stock.

## Privatization and the RHS Prepayment Experience

As early as 1949, Congress recognized the uniqueness of rural America and its housing conditions by creating a separate structure of housing assistance within USDA's Farmers Home Administration (FmHA). (FmHA was reorganized and renamed in 1994, with its housing programs located in a new Rural Housing Service.) This uniqueness is reflected in the Section 515 rural rental housing program. The program was created by the Senior Citizens Housing Act of 1962 to finance the construction, rehabilitation, acquisition, and operation of long-term rental or cooperative housing for lower-income seniors living in rural farm and nonfarm areas. In 1966, it was broadened to include all low- and moderate-income households.

Section 515 was one in the series of private sector subsidy programs created in the 1960s by the Kennedy and Johnson administrations that sought to involve private developers, both profit-motivated and nonprofit, in the production of rental housing.[3] According to Hays (1985, pp. 40-41), these programs, although supported by some conservatives, emerged from the liberal belief that government can positively affect the functioning of the market. Hoping to expand the constituency for social welfare measures and find a positive alternative to the much-maligned public housing program, they encouraged a blended approach of public sector financing and private sector participation.

Like the better known HUD multifamily programs of the time, Section 515 was intended to stimulate rental housing within markets and for populations not normally served by conventional lenders. Unlike the HUD programs, Section 515 was singularly rural in mission as well as method. Financing was provided in the form of *direct* government-subsidized loans (in contrast to the insured loans that HUD provided) with below-market interest rates as low as 1%. Only nonprofit sponsors could receive these loans. For-profit sponsors also could participate in the program but received *insured* nonsubsidized loans at the prevailing government lending rate to produce housing that was of a more moderate- or middle-income character. Direct loans were a response to the reality of severe shortages of private credit and capital in rural areas, resulting in part from the absence of financial institutions and relatively lower incomes of rural people compared with city dwellers. Both direct and insured loans had maturities of up to 50 years.

Section 515 was distinguished from the HUD programs in another critical way. While representing a major new commitment of credit, Congress was careful not to upset precedent first established in the Housing Act of 1949. This precedent was a limitation in RHS's existing Section 502 home-ownership program (then the primary housing resource in rural areas), which mandated that the agency act only as a lender of last resort. Therefore, by statute, borrowers had to satisfy two requirements in the single-family program: a "no credit elsewhere" provision and a "graduation" provision. Under the first provision, loan applicants had to demonstrate that they had insufficient personal resources or that non-RHS credit was either unavailable or available at loan amounts, interest rates, maturities, and other terms that would preclude provision of the housing. Under the second provision, borrowers were required to "proceed with diligence"[4] to graduate to non-RHS credit at the earliest opportunity, whenever such credit could be secured on reasonable terms and conditions.

The practical effect of the graduation requirement was that borrowers, upon RHS request or at their own initiative, could prepay and refinance the balance of Section 515 loans at any time. Where prepayment occurred, not only was the RHS financing retired, but borrowers were released from enforceable agreements in the Section 515 notes or security instruments that enabled RHS to control the use of the housing. For project tenants, this meant the immediate cessation of all government regulation of admissions and rents. Most important, it meant the withdrawal of all government subsidies, both the "shallow" subsidy provided through the low-interest loan and, in some cases, additional "deep" subsidies provided through RHS's rental assistance program.

In this respect, the Section 515 program again differed fundamentally from the HUD programs. RHS's graduation requirement had no parallel in these programs, which imposed a 20-year prepayment restriction on for-profit sponsors and allowed no prepayment option for nonprofit sponsors. Moreover, unlike Section 515 property owners who immediately lost RHS project-based rental assistance upon prepayment, HUD owners still were subject to the terms and conditions of any existing Section 8 rental subsidy contracts that survived the prepayment until the contracts reached the next earliest renewal or expiration dates.

Not until the early 1970s, however, did the graduation requirement emerge as a real threat to the housing and its occupants. During the program's first 10 years, there was little likelihood of prepayment and tenant hardship. Nonprofit borrowers were unlikely to be eligible for private financing that would enable them to graduate and, in any event, were achieving their goal of providing affordable housing. For-profit borrowers already held virtual market-rate loans, and prepayment and replacement with commercial loans would have little impact on tenants' rents. In other words, the program was working essentially as intended—stimulating housing production and providing a dependable housing resource.

This internal symmetry began to change as the result of two administrative decisions made by RHS in the early 1970s. First, in 1971, the agency decided to forcibly reassert its graduation policy because of unacceptably low rates of graduation. According to the Housing Assistance Council (HAC), 1,889 loans were closed from inception of the program through fiscal year 1971, but borrowers of only 57 loans had graduated since the first reported prepayments in fiscal year 1966.[5] (These numbers include both initial and subsequent loans on the same properties, thereby overstating the actual number of projects closed and prepaid.) RHS staff were directed to review the status of active borrowers annually to determine whether changed circumstances could justify requests for graduation. If borrowers refused to graduate voluntarily, then legal action was to be considered. To ensure that this policy was carried out properly, district supervisors were ordered to conduct spot checks at the county office level, and state directors were required to report to the national office on progress made in the graduation of borrowers.

Second, in 1972, RHS adopted rule changes permitting limited-profit sponsors—typically, nonsyndicated profit entities organized as individuals, partnerships, or corporations and syndicated limited partnerships—to receive subsidized loans. In turn, they had to agree to methods of operation and income controls for tenant admissions and rental charges that would ensure occupancy by low- and moderate-income households. They also had to accept a lower return on their initial investments, that is, limited profits that were less than what could be realized with RHS nonsubsidized loans, which permitted operation on a full-profit basis. The return was set at 6% per annum (later increased to 8%).

With this second decision, the character of the Section 515 program was substantially altered. Production skyrocketed as intended. From fiscal years 1973 to 1979, Section 515 was transformed from a comparatively low-production program used mostly by nonprofit and small "mom and pop"-type operators of rental housing to one dominated by syndicates of profit-motivated investors (FmHA, 1979). Table 10.1 shows that, in August 1979, limited partnerships and other limited-profit entities accounted for just under half of all loans and nearly 60% of all units and dollars, whereas nonprofit sponsors accounted for about one quarter of the caseload. The number of loans nearly quadrupled, and loan dollars rose more than 2,500% over the 10-year period from fiscal years 1963 to 1972 (Table 10.2) (FmHA, 1981).

The inclusion of limited-profit sponsors, however, resulted in a peculiar contradiction. While seeking greater production, RHS still was mandated to encourage and compel prepayments in accordance with its 1971 instructions but without consideration of the tragic consequences for the very low-income occupants the subsidized loans were intended to help. Unlike before, prepay-

**TABLE 10.1**  Section 515 Caseload by Sponsor Type as of August 20, 1979

| Sponsor Type | Number of Loans | Number of Units | Dollar Amount |
| --- | --- | --- | --- |
| Full profit | 1,524 | 16,792 | 222,083,999 |
| Public body | 243 | 6,550 | 140,159,646 |
| Nonprofit | 1,841 | 30,812 | 500,516,343 |
| Limited partnership | 1,026 | 33,008 | 604,933,122 |
| Other limited profit | 2,240 | 40,726 | 671,782,250 |
| Totals | 6,874 | 127,888 | 2,139,375,360 |

SOURCE: Farmers Home Administration (1979).

ment of these loans by limited-profit borrowers was not only economically feasible but also desirable once the main incentive to limited-profit investment—the tax-sheltered depreciation benefits—had expired, usually after 10 years. When coupled with considerable value appreciation, the attractions of prepayment and conversion to market-rate operations were unmistakable. In essence, RHS opened the front door to limited-profit borrowers without closing the back door. No other federal housing program offered borrowers both the advantages of subsidized financing and of quick egress from the obligations imposed by the financing.

By the end of the decade, loan prepayments were on the rise. Litigation in Ohio and reports of rent increases, constructive eviction of tenants, and loss of the housing in a few publicized cases gave a human face to the issue. High property appreciation in certain rural parts of the country, and the growing number of limited-profit sponsors that could be expected to reach their tax shelter "burnouts," augured an even larger wave of prepayments in the 1980s.

With this threat in mind, the first-ever anti-prepayment legislation was enacted on December 21, 1979, after a prolonged, highly ideological debate that kept Congress at a "sword's edge" (AuCoin, 1979, p. 36913) for months. Owner advocates and conservatives in Congress argued that owners' property rights were paramount—that retroactivity was a breach of contract and unconstitutional. Tenant advocates and their liberal supporters countered that prepayment violated congressional intent and the public purpose of providing long-term, affordable housing (Wiener, 1993). The legislation retroactively restricted prepayment of subsidized loans for a period of 20 years from

**TABLE 10.2** Section 515 Rural Rental Housing Loans for Fiscal Years 1963 to 1979

| Fiscal Year | Number of Loans | Dollar Amount |
|---|---|---|
| 1963 | 2 | 117,000 |
| 1964 | 21 | 1,166,600 |
| 1965 | 32 | 1,787,530 |
| 1966 | 76 | 4,165,770 |
| 1967 | 99 | 5,314,350 |
| 1968 | 276 | 12,427,010 |
| 1969 | 329 | 15,351,280 |
| 1970 | 412 | 24,148,230 |
| 1971 | 325 | 21,012,400 |
| 1972 | 372 | 21,502,910 |
| 1973 | 546 | 87,886,880 |
| 1974 | 562 | 140,663,110 |
| 1975 | 903 | 267,036,590 |
| 1976 | 685 | 213,992,470 |
| 1977 | 1,114 | 518,017,210 |
| 1978 | 1,056 | 627,442,260 |
| 1979 | 1,247 | 827,894,470 |
| Totals | 8,057 | 2,789,908,070 |

SOURCE: Farmers Home Administration (1981).
NOTE: Table includes direct loan costs and excludes insured loan costs.

the dates the loans were approved (15 years for nonsubsidized loans). It extended the same restrictive use to prospective projects. The measure was fully supported by RHS, which wrote that it had acted without sufficient foresight in 1972 when it authorized provision of subsidized loans to limited-profit sponsors.[6]

The following year, however, strenuous opposition from the building industry resulted in repeal. The 20-year prepayment restriction survived for loans approved after December 21, 1979, the effective date of the 1979 legislation. Borrowers of these "post-1979" loans could prepay before 20 years only when (a) they agreed to keep the units affordable for the balance of the restrictive-use period, (b) RHS determined that there no longer was a need for the housing, or (c) federal or other financial assistance no longer was available. Borrowers of "pre-1979" loans, however, once again were free to prepay without restriction. As of August 1991, 110,383 units fell into the pre-1979 category out of approximately 360,000 Section 515 units nationwide.[7]

During subsequent years, the pace of prepayment quickened. According to GAO (1988), RHS records and prepayment reports showed that the agency had approved loan prepayments on 408 projects from inception of the Section 515 program until February 1987. Nearly 90% of the projects and 5,557 units were prepaid since fiscal year 1984 (Table 10.3).

Concern about tenant displacement, including information from prepayments in California, New York, Washington, and other states, forced Congress to revisit the issue in 1986, a year that saw a record number of prepayments. A moratorium was placed on all prepayments, commencing on October 18, 1986, and ending on June 30, 1987. The moratorium remained in effect, with a few short gaps, until March 15, 1988, when it was succeeded by a "permanent" preservation scheme adopted in the Emergency Low-Income Housing Preservation Act (ELIHPA) of 1987. Interim regulations implementing the act became effective in May 1988.

ELIHPA established strict standards for approving prepayment requests and a formal process for preserving pre-1979 housing. First, RHS was directed to offer financial incentives to owners requesting prepayment to encourage them to stay in the program for at least 20 more years. The main incentives were a 90% equity loan and a greater rate of return on investment. To ensure continuing affordability to project tenants, owners could receive interest rate reductions on the existing loans as well as incremental rental assistance. Second, when owners refused "stay-in" incentives, RHS could approve prepayment only if there would be no adverse impacts on the tenants, the local supply of affordable housing, or minority housing opportunities. Finally, in cases of projects not able to satisfy the prepayment exceptions, RHS was required to force sales to qualified nonprofit or public agencies within 180 days through provision of acquisition loans for 102% of the owners' equity. A maximum of 5,000 units could be financed each year.

## Programmatic and Legislative Developments Since ELIHPA

Since adoption of ELIHPA, national data on the number of projects prepaying or receiving incentives to sell or stay in the Section 515 program are spotty at best. In violation of congressional intent, RHS has failed to date to

**TABLE 10.3**  Section 515 Rural Rental Housing Loans Prepaid From Inception to February 1987

| Fiscal Year | Number of Projects | Number of Units |
|---|---|---|
| Prior to fiscal year 1984 | 53 | — |
| 1984 | 68 | 961 |
| 1985 | 86 | 1,364 |
| 1986 | 176 | 2,756 |
| 1987 | 25 | 476 |
| Totals | 408 | 5,557 |

SOURCE: U.S. General Accounting Office (1988).

create a national preservation office with permanent staff and has not established a system of centralized administration and data tracking. This has made it very difficult for objective external assessment of the successes and shortcomings of the law and RHS implementation.

The most comprehensive performance evaluation of the preservation program was undertaken in 1992 by a National Task Force on Rural Housing Preservation (1992) convened by HAC, the California Coalition for Rural Housing Project, and National Housing Law Project with funding from the Ford Foundation. At the time of the evaluation, RHS had received notices of intent to prepay from owners of more than 300 projects and had offered incentives to at least 136 projects, nearly all of which were owner stay-ins. Among other conclusions, the task force found that, overall, ELIHPA was achieving the goal of preventing the wholesale prepayment and loss of pre-1979 housing. Using a model to simulate owner choices, it predicted that the majority of owners would prepay in the absence of the law because they held properties that could support conversions to conventional market rents. The after-tax compensation provided to owners in ELIHPA was, on average, worth somewhat less than the after-tax value of their market alternatives.

In many cases, however, the task force found that RHS was offering the maximum incentives to properties that appeared to have little market justification for conversion or an equivalent or only marginally more profitable conventional use. For example, 90% equity loans were routinely approved for stay-ins when RHS had discretion to offer smaller loans to convince owners of less economic properties to extend affordability. The "richness" of these incentives, together with the fact that cash from equity loans was not a taxable transaction while most sales had a high capital gains tax cost, greatly undercut the sales component of ELIHPA. Moreover, these loans were inconsistent with the requirement that RHS pursue the least costly alternative to the federal government. In exchange for incentives, RHS was leveraging only the bare minimum 20 years of additional restrictive use required when it had discretion to negotiate much longer affordability.

The task force also reviewed RHS administration of the prepayment exceptions in the case of 41 pre-1979 properties. It concluded that most prepayments did not conform with the prepayment exceptions or were approved without observing the procedural scheme in ELIHPA—offering incentives to stay in or forcing sales to nonprofit entities. Different RHS state and district offices were using different criteria and standards for prepayment approval and variable verification and documentation methods. For example, findings on possible tenant displacement, the availability of alternative housing, and loss of minority housing opportunities were incomplete and unsupported. RHS controls for ensuring continued affordability of projects prepaying prior to the end of the 20-year restrictive-use period were wholly inadequate. In a few instances, the agency was using prepayment as a servicing action to accelerate loans of bad owners and to effectively remove them, along with their properties, from the Section 515 program.

Following precedent set in ELIHPA, preservation of post-1979 projects was addressed in subsequent legislation (Anders, 1993). The HUD Reform Act of 1989 imposed a 50-year restrictive-use period coterminous with the full mortgage term for all projects with loans approved after December 15, 1989. Legislation adopted in 1993 extended the preservation scheme in ELIHPA to all projects approved after December 21, 1979, and before December 15, 1989—a total of 256,211 units (National Task Force on Rural Housing Preservation, 1992, p. 36). At the end of the 20-year restrictive-use period created in the 1979 legislation, these projects were to be subject to the same prepayment exceptions and preservation financial incentives as were pre-1979 projects. The legislation also directed RHS to create an Office of Rural Housing Preservation to centralize program administration, monitoring, and data collection.

In recent years, obligations for RHS's preservation program have been hit by deep cuts in appropriations for the overall Section 515 program. Prior to fiscal year 1995, about $20 million annually was spent on the provision of preservation financial incentives, primarily equity loans for stay-ins. In the 3-year period from fiscal years 1995 to 1997, RHS obligated just over $9 million for equity loans on about 40 projects,[8] only a small portion of the total projects seeking incentives.

Although the ELIHPA prepayment restrictions and financing scheme have not endured the scrutiny and challenges that led to the cutoff of funds for HUD's preservation component in 1997, ominous signs are on the horizon. Court challenges, decreases in Section 515 program spending, and efforts by RHS to find alternative funding mechanisms to limit the budget impacts of the preservation program might have the effect, in the not-too-distant future, of undermining the program. Once again, the majority of pre-1979 projects could be exposed to prepayment, reduced affordability, and tenant displacement.

## Conclusion

The legislative, regulatory, and administrative origins of the prepayment problem in RHS's Section 515 program are unique. The ideological roots of the problem, however, can be found within the policy framework of U.S. housing history. During the first 25 years of the federal housing program, publicly owned and managed housing was the main way in which the government satisfied the shelter needs of poor Americans. Both Section 515 and the HUD multifamily programs of the 1960s were predicated on the correct assumption that private, profit-motivated owners could be enticed to build, own, and operate low-income rental housing in large numbers. In both instances, however, there was little, if any, consideration of the consequences for tenants or the long-term preservation of the housing when investor incentives to remain in the programs ran out.

If anything, history shows that prepayment is wholly consistent with the logic of the private sector subsidy strategy that emerged during this period in U.S. housing policy. The central premise of the new strategy was that the need to expand the supply of affordable accommodations for more moderate-income families was best met by creating ever more lucrative opportunities for profit-motivated entities. With each program innovation, the production process became dependent on for-profit actors within an increasingly privatized delivery system comprising private lenders, investors, builders, and others. Prepayment was only the most advanced form of privatization.

Now, the flawed liberal policies of the 1960s have come full circle, as hundreds of thousands of units are at risk of conversion nationwide. In 1996, the 104th Congress removed the prepayment restrictions first imposed on HUD rental housing by ELIHPA and later by the Low-Income Housing Preservation and Resident Homeownership Act (LIHPRHA) of 1990. In 1997, the 105th Congress effectively ended the sales program in LIHPRHA that enabled tenant groups and tenant-endorsed nonprofit entities to acquire the housing as an alternative to prepayment. Rather than recognize the enormous public investment in this housing and the current and future needs of lower-income families, the Clinton administration and its allies in Congress have pursued a nonpreservation course that will permanently and irrevocably dismantle this vital national asset.

The same economic rationales that motivated RHS owners to prepay properties in the 1980s can be expected to operate in the HUD inventory in the 1990s and beyond as financial, tax, and market conditions permit. The prepayment experience in the Section 515 program should give pause to those who would assume, all too casually, that the current project-based system can be transformed seamlessly to a tenant-based system without short- or long-term harm to tenants. It also should point out the futility of continuing a policy that places the future affordability and disposition of low-income housing in the hands of for-profit owners.

# Notes

1. The provision of market-rate financial incentives as compensation for restricting owner prepayment rights, and to convince owners to extend project affordability, originally was supported by some conservatives who realized that Congress would not permit unilateral prepayment. Moreover, owners understood that the incentives, in many cases, would be more lucrative than the alternative conventional uses of the property upon prepayment.

2. In March 1996, Congress repealed prepayment restrictions imposed in 1987 that protected more than 400,000 HUD-subsidized units developed under its Section 221(d)(3) below-market interest rate and Section 236 programs. In October 1997, Congress "zeroed out" fiscal year 1998 appropriations for HUD's preservation program created in Title VI of the National Affordable Housing Act of 1990.

3. In 1961, Congress created the first of the HUD mortgage insurance programs for multifamily housing, Section 221(d)(3). This was followed in succession by the Section 221(d)(3) below-market interest rate program in 1965 and the Section 236 program in 1968.

4. 42 U.S.C.A. § 1472(b)(3).

5. See Housing Assistance Council (1987, Appendixes F and H-2). This report was released prior to the 1988 GAO study cited earlier, which showed a much smaller number of prepaid loans. According to HAC, the nearly 1,300 loans reported as prepaid by RHS, in its report up to fiscal year 1986, overrepresented the total projects prepaid because the computer-generated count

contained an unknown number of initial and subsequent loans on the same properties. Thus, some projects were double-counted, whereas others may have prepaid subsequent loans, but initial loans remained.

6. Letter from Gordon Cavanaugh, administrator, FmHA, to Congressman Les AuCoin, author of amendment to retroactively limit prepayment, Washington, D.C., 1979.

7. Precise counts of the number of pre-1979 units subject to unrestricted prepayment have proven surprisingly elusive over the years. The total indicated by sponsor type in Table 10.1 as of August 20, 1979, was 127,888. A survey of pre-1979 units by RHS in September 1986 showed a total of 159,534 units out of 357,318 Section 515 units nationally. The GAO report, using data from RHS's MISTR computer system from February 1987, counted 124,834 pre-1979 units (6,626 projects) out of 360,749 total units (17,162 projects). The 110,383-unit number is from an August 1, 1991, computer run by RHS from its newer AMAS computer system, which in all likelihood does not double-count properties and is the most accurate.

8. RHS, Multifamily Housing Processing Division, year-end reports for fiscal years 1995 to 1997, Washington, D.C., 1997.

# Chapter 11

# Credit and Capital Needs for Affordable Rural Housing

*Leslie R. Strauss*

Improving access to financing is an essential part of improving housing in rural America. Credit (loans) and capital (investment) must be available to developers of new housing at a cost that enables them to produce units affordable to those in need. Affordable credit must be available for lower-income households to buy homes and to repair their homes and for organizations to purchase and repair multifamily housing.

It is nearly impossible to determine how much credit presently is available in rural areas and how much is needed. Although recent studies indicate that the rural credit and capital shortage is not as severe as it once was (Economic Research Service, 1995; Kennedy, Goldberg, & Fishbein, 1990), all indications are that available financing falls far short of meeting the need. Nonmetropolitan borrowers as a whole must make higher down payments, accept shorter loan terms, and pay higher interest rates than those in metropolitan areas.

Credit problems are much worse in some rural places than in others. One study noted that often the worst "pockets" were "in somewhat remote areas of the country or in communities in which minorities or lower income residents predominate" (Kennedy et al., 1990, p. 1). Another found that although the "typical" rural county is now served by five bank firms (which may be more than five offices if a firm has multiple branches), 20% of rural counties have two or fewer bank firms, and most of those counties are in low-population areas such as the mountain states and the High Plains or in areas with persistent concentrations of poverty such as the Mississippi Delta, Appalachia, and Indian reservations (Economic Research Service, 1995, p. 2). These findings are consistent with the Housing Assistance Council's (HAC) observation, based on more than 25 years of experience with low-income rural housing, that smaller, more remote areas are more likely than larger communities to lack access to mortgage credit (Loza, 1994, p. 4).

Some part of this rural credit gap arises because rural economies and properties often do not meet urban-oriented lending standards. The economy of a rural area is more likely to rely on a single industry, making lending more risky. Properties often do not meet conventional underwriting requirements because they might have larger sites than comparable urban properties, be located in areas that are less than 25% developed, have numerous outbuildings, combine business and residential use, be subject to unusual easements, have incomplete or no utility service, and/or be located on unpaved roads (Kennedy et al., 1990, p. 6; U.S. General Accounting Office [GAO], 1991). It might be difficult or impossible to obtain appraisals based on recent sales of comparable homes in the area because often there are no such sales, and in very remote areas there might be very few homes in the vicinity.

## Single-Family/Homeownership Mortgage Financing

Little quantitative information exists about the adequacy of the housing-related credit currently available in rural areas. The Home Mortgage Disclosure Act (HMDA), which requires medium-sized and large lenders in metropolitan areas to make information available about the characteristics of persons receiving and being denied loans, has provided a rich source of data for research on credit gaps in large cities that, until recently, was not available to nonmetropolitan researchers. Unfortunately, by contributing to the tendency of researchers and the media to focus on urban problems, the rural information gap resulting from HMDA might have reinforced a perception that rural areas have no credit problems.

The existence of a shortage of mortgage credit in rural areas is dramatically illustrated by the fact that 7% of home purchasers in nonmetro areas, but only 3% of such purchasers in metro areas, obtain first mortgages from individuals (U.S. Bureau of the Census, 1994). It is unlikely that nonmetro sellers' preference for financing sales is twice as strong as that of metropolitan sellers. More nonmetro buyers simply have no other source of credit available.

Those rural residents who can get mortgage credit must pay more for it because the mortgages available in rural areas tend to have higher interest rates and shorter amortization periods than those in cities. In 1995, the median interest rate for nonmetro mortgages (including government-subsidized ones) on owner-occupied homes was 8.7%, compared with 8.3% for central city borrowers and 8.2% for suburban homeowners. The median term of primary mortgages in nonmetropolitan areas was 23 years, compared with 30 years in central cities and 29 years in suburbs (U.S. Bureau of the Census & U.S. Department of Housing and Urban Development [HUD], 1997, Table 3-15). In case studies conducted around the country, one study found that mortgage terms varied widely, from the same as those available in competitive metro markets to a highly onerous requirement for a 30% down payment for a 5-year mortgage purchase loan (Kennedy et al., 1990).

Nonmetro borrowers might have to provide larger down payments as well. One study found that private mortgage insurance was not available in several nonmetro communities and that where private mortgage insurance was not available, buyers were required to put at least 20% down to get mortgage loans from banks or savings and loan associations (S&Ls). In those areas, therefore, borrowers—particularly those with low incomes— were much less likely to obtain private lender financing (Kennedy et al., 1990, p. 7).

The insufficiency of rural mortgage credit also is indicated by existing housing conditions. A shortage of affordable home equity loans for rehabilitation of housing, both owner occupied and rented, might help explain the fact that a greater proportion of housing is physically deficient in rural areas than in cities.

In addition, the remarkably high proportion of mobile homes[1] in rural and nonmetro areas might be an indication that insufficient mortgage credit is available. Fully 17% of rural homeowners in 1990 owned mobile homes or trailers, compared with 8% nationwide[2] (U.S. Bureau of the Census, 1992a, Census of Population and Housing, 1990, Summary Tape File 3), not necessarily because they preferred those units but rather because such units are less expensive and easier to finance than site-built homes. A mobile/ manufactured home purchased from a dealer usually is financed with a personal property loan (the same type of consumer loan that finances a car) rather than a mortgage. For those who can buy new mobile/manufactured homes, such loans are easier to obtain than mortgages, although borrowers do not build equity in their homes and receive none of the tax advantages of purchasing real estate. Also, despite the cost advantages of mobile homes, in 1995, about one quarter of nonmetro households living in mobile homes— the same proportion as those occupying non-mobile units—were cost burdened; that is, they paid more than 30% of their incomes for housing costs (U.S. Bureau of the Census & HUD, 1997, Table 10-13).

Apparently, nonmetro buyers would not prefer to own manufactured homes if financing and costs were the same. A study of consumer preferences in the South, where manufactured homes are common, found that respondents preferred conventional homes to mobile homes (Shelton & Sweaney, 1983). Similarly, although manufactured homes are more common in rural areas, another study found no real differences between housing preferences in urban and rural areas. Observed differences were attributable instead to differences in housing prices and household income (Copley International Corporation, 1979). It is likely that the lower purchase prices of manufactured homes, and the lack of mortgage credit that would make the purchases of conventional homes affordable, combine in rural areas to make "trailers" a feasible selection for lower-income purchasers.

The high rate of ownership of mobile homes in nonmetro areas might be one reason why nonmetro homeownership rates are high despite mortgage credit problems. High ownership rates also could be due, at least in some areas, to the availability of low-cost family-owned land, low turnover of units, and/or the presence of second homes owned by relatively wealthy persons.

## Multifamily and Rental Housing Financing

Provision of mortgage credit for multifamily and rental housing in nonmetro areas is hampered by the same problems as provision of single-family credit. In addition, because rural multifamily projects generally are smaller than those in cities, lenders (like developers) who serve mixed geographic areas and can choose urban projects might do so to maximize profit. However, even fewer data are available about financing the development of multifamily housing in nonmetro areas than about financing single-family home purchases. Residential Finance Survey data indicate that rental housing in rural areas, like owner-occupied housing, is financed primarily by banks. Only 4% of nonmetro rental housing is sold to secondary market agencies such as

Fannie Mae and Freddie Mac—a much lower proportion than in metro areas (U.S. Bureau of the Census, 1994).

In all geographic areas, the secondary market agencies purchase fewer multifamily (primarily rental) mortgages than single-family mortgages. Multifamily mortgages, particularly those for affordable housing, are harder to securitize because they are less standardized, they have more volatile cash flows, and there is less information about their performance. Whereas these characteristics make sales in the secondary market more difficult, it also is troublesome for rural lenders to keep multifamily loans in their own "portfolios" because capital standards imposed by the Financial Institutions Reform, Recovery, and Enforcement Act of 1989 discourage portfolio lending (Office of Policy Development and Research, 1995).

Multifamily housing credit was further tightened in the early 1990s by the ongoing aftereffects of the S&L problems of the mid- and late 1980s. Because they are relatively risky and volatile, multifamily loans, especially for affordable housing, fit the category that lenders, insurers, and the secondary market want to avoid. As a result, multifamily credit became harder to obtain in all geographic areas than it was a decade earlier (Hamilton Securities Group, National Multi-Housing Council, & National Apartment Association, n.d.; Office of Policy Development and Research, 1995). Recognizing an unmet need and an untapped source of revenue, however, lenders are returning to multifamily lending. Freddie Mac, after years of inactivity with respect to multifamily loans, began purchasing them again in late 1994.

## Special Problems

Persons of color living in rural/nonmetro areas suffer disproportionately severe housing problems and persistent poverty. They suffer mortgage credit shortages as well, but there is little research on this topic.

A 1977 report on equal housing opportunity did include some relevant findings on access to mortgage credit. African American and Hispanic residents of many of the 12 rural communities studied believed minorities had less access to credit than did white persons. Minority respondents stated that they faced barriers to obtaining information about conventional financing and that financial institutions treated minority loan applicants differently from whites (Urban Systems Research & Engineering, 1977). Although researchers found it difficult to document these problems objectively and loan officers denied that discrimination existed, more recent analyses of

HMDA data provide documentation for the existence of differential treatment in the loan application process in major metropolitan areas (Carr & Megbolugbe, 1993; Munnell, Browne, McEneaney, & Tootell, 1992). It is likely that these disparities exist in nonmetro places as well given that there is no reason to believe that nonmetro areas have made more progress in nondiscrimination since 1977 than have cities.

In addition, Hispanic residents of the U.S.-Mexico border region experience a particular type of mortgage credit problem. Many of them live in *colonias,* small communities in which lots often were sold without being subject to legal restrictions and in which the sellers often provided purchase financing through "contracts for deed" instead of mortgages. This type of financing does not provide a public record of a buyer's purchase. It allows the seller/lender to retain title to the property until the debt is fully paid and, therefore, to repossess the lot (and whatever the purchaser has built on it) immediately if even a single payment is missed. The "homeowner" does not have any equity in the property and cannot use it as security for a loan.

Native Americans, who as a group live in the worst rural housing in the country, face another type of rural mortgage credit shortage. Because native lands are held in trust and, therefore, cannot be taken by foreclosure in the event of mortgage default, standard mortgages effectively have not been available to rural Native Americans for either single-family or multifamily housing (McLaughlin, 1993, p. 2; National Commission on American Indian, Alaska Native, and Native Hawaiian Housing, 1993, p. 9). Native areas also suffer from extreme versions of problems present in other rural places. The market values of homes tend to be well below the homes' construction or replacement costs (McLaughlin, 1993, p. 2), and infrastructure such as roads and utilities often is incomplete.

## Rural Lenders

Differences in the character of the available lenders account for some of the rural/urban differences in mortgage credit. There are proportionately fewer S&Ls in rural areas than in cities, and there are very few mortgage bankers in rural areas. Although the recent trend toward consolidation of the banking industry has meant that some rural banks are now branches or subsidiaries of large regional banks, nonmetro banks and S&Ls still tend to be smaller than their metropolitan counterparts. Slightly more than half of the nearly 11,000 commercial banks insured by the Federal Deposit Insurance Corporation (FDIC) in 1993 had headquarters in nonmetro communities, but they held only 14% of U.S. bank assets. Many more banks than thrifts had headquarters in nonmetro areas—in 1988, about eight times more ("The Structure of the Banking and Thrift Industry in Rural Areas," 1991). Since that date, the proportion has increased; approximately 10% of nonmetro thrifts disappeared between 1989 and 1991, albeit "with negligible disruption of financial markets" (Rossi, 1991, pp. 36-37).

Mergers or affiliations with large interstate holding companies seem to have both positive and negative effects for rural borrowers. Large lenders may have a wider range of products, but decisions might need to be made somewhere far away from the rural office (Milkove & Sullivan, 1990). Mergers also may lead to the closing of small, relatively unprofitable branches.

Where small rural lenders still exist, they have smaller and, therefore, less specialized staffs. Rural lenders, particularly those in remote areas, tend to have less technical capacity to evaluate risk and undertake complex transactions, complicated by a lack of capacity on the part of borrowers to develop sound applications. Some rural lenders still prefer to base their lending decisions on what they know about their applicants rather than on standardized criteria. Rural lenders also are more conservative than those in cities, maintaining higher equity capital ratios and investing more often in safe government securities than in local loans. Metropolitan lenders lend out a higher proportion of their funds (Kennedy et al., 1990, p. 8; Mikesell, 1985; Office of the Comptroller of the Currency, 1992).

Similarly, rural lenders use the secondary market less often than do their urban counterparts, and the former are more likely to hold loans in portfolio. According to the U.S. General Accounting Office (1991), two reasons for this difference are that

> (1) the low volume of loans [that rural lenders] originate is not attractive to home mortgage secondary markets and does not warrant the amount of resources the bank must dedicate to participating in one, and (2) the lack of understanding of secondary markets and the administrative burden they create is a disincentive for banks to participate. (p. 30)

The Federal Agricultural Mortgage Corporation (Farmer Mac) was created in 1988 to resolve some of these problems, but it serves only places with less than 2,500 residents and generally has been inactive in housing. Thus, potentially important sources of increased mortgage finance in rural areas are the government-sponsored enterprises that comprise the national secondary market—Fannie Mae and Freddie Mac—as well as the Government National Mortgage Association (Ginnie Mae), a part of HUD.

A variety of federal programs and incentives exist to help lenders make housing credit available nationwide, and these could be better used in rural areas. For example, the Federal Housing Administration (FHA) has been insuring privately issued mortgages since the 1930s and now serves home-buyers not covered by private insurers. Yet, nonmetro lenders have been less likely than metro lenders to use FHA programs, so urban homeowners are more than twice as likely to have FHA mortgages as their rural counterparts (Office of Policy Development and Research, 1995, p. 3; U.S. Bureau of the Census & HUD, 1997, Table 3-15).

A recent data review indicated that in metro areas, "borrowers and lenders use FHA insurance as the instrument of choice to extend home mortgage credit to underserved groups and markets" (Office of Policy Development and Research, 1995, p. 3). It serves this purpose to some extent in nonmetro areas as well, at least for some portions of the population. FHA provides a far greater proportion of mortgages for African American and Hispanic households than for the population at large in both metro and nonmetro areas. In nonmetro areas, 17% of mortgages on owner-occupied homes with African American householders, and 16% of mortgages with Hispanic householders, are FHA insured (U.S. Bureau of the Census & HUD, 1997, Tables 5-15 and 6-15).

Research indicates that the reasons for low nonmetro usage of FHA mortgages are connected to the nature of nonmetro lending institutions (U.S. General Accounting Office, 1981; Housing Assistance Council, 1981, 1985, 1986a; National Association of Home Builders, 1992; Pheabus & Rowland, n.d.). Many have too little capital to be approved as FHA lenders, and the low volume of mortgage loans in some rural areas combines with the low FHA loan fees to make the program uneconomical for some banks. Small lenders interviewed by researchers repeatedly asserted their belief that FHA and other "government programs" require too much paperwork and red tape. Even when they qualify or might be interested, small rural financial institutions seldom have the administrative resources to participate. Some blame FHA itself, charging that the agency lacks interest in rural areas.

Other federal programs to encourage private lending for affordable housing developed along with increased regulation of lenders after the S&L crisis in the late 1980s. The Community Reinvestment Act (CRA), first enacted by Congress in 1977 to combat redlining, has received the most publicity. CRA's usefulness for advocates of housing for low-income persons was substantially increased in 1989 when a statutory amendment and new regulations required federal regulators to make lenders' CRA performance ratings public and to take them into account when deciding whether to approve lenders' requests to merge or to close or open new branches. Thus, lenders have had to justify to their communities (urban and rural) their records of lending for affordable housing, consumer credit, small businesses, and small farms. Although advocates have not always agreed with regulators' assessments of individual lenders, community groups and regulators have been able to use CRA as both a carrot and a stick to increase private lending for community needs including affordable housing.

Other "carrot and stick" programs for federally regulated lenders include the Affordable Housing Program (AHP) and Community Investment Program (CIP) of the Federal Home Loan Bank (FHLB) system. Both involve subsidies from Federal Home Loan district banks to lending institutions that are members of the FHLB system, thus encouraging member institutions to make loans for affordable housing.

The Rural Housing Service (RHS), an agency of the U.S. Department of Agriculture, bridges part of the rural mortgage credit gap but not all of it. RHS, which administers federally financed rural housing programs formerly under the jurisdiction of the Farmers Home Administration (FmHA), is intended to be a lender of last resort. Among other programs, RHS provides single-family mortgages to purchasing households and permanent loans for developers of multifamily housing. These rural housing programs are extremely successful, but demand has consistently exceeded available funds; the programs have not been large enough to meet the need. Also, in the mid- and late 1990s, Congress has shifted RHS funds from direct mortgage lending to guaranteeing mortgage loans made by private lenders. Guarantee programs cost the government less but tend to serve moderate-income clients rather than those with low incomes because borrowers must meet the qualifications established by private lenders and must be able to pay market interest rates.

Rural mortgage credit for both owner-occupied and rental housing is provided through other federal programs as well. HUD provides funds through a number of programs; the Community Development Block Grant and HOME programs are particularly important in rural areas. The Department of Veterans Affairs (VA) provides another government guarantee program; VA loan guarantees are available to veterans, regardless of income level or geographic area, and require them to pay some fees but not to provide down payments. Through another program, HUD guarantees private mortgage loans for homes on Native American trust lands.

Farm Credit System (FCS) lenders provide an additional source of purchase money mortgages for moderate-income borrowers living in open country or small towns of up to 2,500 residents and not associated with larger population centers, although FCS is not a major source of housing credit even for its moderate-income constituency. In fiscal year 1990, FCS lenders held 1.3% of the outstanding mortgages in rural areas of under 2,500 residents and did not approach the 15% FCS cap (U.S. General Accounting Office, 1991).

Another government program that helps finance multifamily housing development in rural areas, as in cities, is the Low Income Housing Tax Credit. In exchange for an investment in the development or substantial rehabilitation of rental housing for low-income residents, an individual or a corporation receives a reduction in the dollar amount of federal taxes owed. The amount of tax reduced is tied directly to the proportion of low-income persons among the residents.

State and local governments, in addition to nongovernmental funding sources, also provide some assistance for housing development. State housing finance agencies are particularly important because they sell tax-exempt bonds to obtain funds for mortgage financing and also administer federal block grant funds. Other examples include foundation grants to nonprofit developers and HAC's seed money loans to local development organizations. However, these sources comprise a very small proportion of the total credit used for rural housing.

## Recommendations

The information presented in this chapter has demonstrated that the private mortgage sector is not providing sufficient mortgage credit for low- and moderate-income residents of rural areas. Federal programs have bridged some part of, but have not closed, the rural credit gap. As the federal role in social programs is further reduced, this gap can only become more pronounced.

"Leveraging," "partnerships," and "guarantees" seem to be the credit provision buzzwords of the 1990s. Developers of affordable housing already are familiar with the need to fit together multiple sources of financing to make a single project feasible and with the fact that, once one source has promised its assistance, its commitment can be used to leverage other funds. Some funding sources, such as the HOME program, require such leveraging. At the same time, to increase their impact, government subsidy programs increasingly encourage public-private partnerships. As part of the effort to reduce government financing for housing, some lawmakers now look to government guarantees as replacements for direct subsidy programs.

These are not complete solutions. There are limits to the amounts that can be leveraged and the number of partners that can be found, particularly in economically depressed communities where the housing need might be the greatest and local resources the smallest. Even with guarantees, private lenders cannot provide all the mortgage credit needed and still meet their requirements for profitability and low risk. It seems that the federal role must continue, and even increase, to house rural residents properly. There is no inexpensive or easy way in which to house everyone properly.

Thus, the housing credit programs of the future must include the same mix of subsidies as those currently existing, from direct low-interest loans for both individuals and developers to guaranteed loans, secondary market purchase requirements, incentives for private lenders, and more. Efforts to improve many of the components of this mix are under way:

- HUD and lawmakers are studying ways in which to reform HUD and its delivery system including FHA.
- HUD has promulgated regulations requiring the government-sponsored enterprises it regulates (Fannie Mae and Freddie Mac) to make specified portions of their secondary market purchases in underserved rural and urban areas.
- New Community Reinvestment Act regulations began to take effect on January 1, 1996, intended to focus on lender performance rather than on paperwork (a charge often leveled by critics of the previous regulations).
- Lenders, both public and private, are strengthening their efforts to work with borrowers to avoid possible problems. Homebuyer education programs, credit counseling, and delinquency and default counseling often are provided, and even required, for low- or moderate-income home purchasers.
- Some private lenders have created community development corporations to extend more credit for community purposes such as mortgage lending without affecting the way in which the lenders do business. Although these entities cannot substitute for lending efforts by banks and S&Ls themselves, they are useful additions.

Additional steps could be taken as well:

→ A recent study recommended national standardization of mortgage provisions and documents along with local flexibility as essential to achieving better financing mechanisms for affordable multifamily housing (National Task Force on Financing Affordable Housing, 1992).

→ Rural lenders should be more willing to provide certain types of loans they often view with skepticism—construction loans, lines of credit, guaranteed loans, and predevelopment loans. All of these could be increased within typical safety and soundness margins.

→ FHA should make an effort to increase its presence in rural areas including educating lenders about its use and assisting them in dealing with the paperwork required.

→ FHA, Fannie Mae, and Freddie Mac should make special rules and arrangements for rural banks so that more of them can become approved lenders.

→ HMDA should apply to all lenders, not just large metropolitan area ones, and should require them to provide separate data for each branch rather than lumping all the data together and attributing them to the geographic location of their headquarters.

→ As the banking industry consolidates, special protections might be needed for residents of areas served by small local lenders purchased by large institutions with headquarters elsewhere so that closures of small branches do not leave rural residents without access to credit on reasonable terms.

→ Government institutions must develop, and encourage private lenders to develop, specialized products that serve the needs of rural borrowers. Such products might range from creation of lending institutions dedicated to providing mortgage credit in Native American areas where land is held in trust to provision of mortgages that replace contracts for deed in colonias. They might include secondary market purchases of smaller than standard mortgages on which low-income rural residents have good payment histories.

# Notes

1. The term *mobile home,* rather than *manufactured housing,* is used here because it is the term used by the Bureau of the Census in compiling the data cited here.

2. See U.S. Bureau of the Census (1992a), Census of Population and Housing, 1990: Summary Tape File 3 on CD-ROM (name of state). [Machine-readable data files]. Calculations are by HAC.

For additional information concerning the files, contact Data User Services Division, Customer Services Branch, Bureau of the Census, Washington, DC 20233. Telephone: (301) 763-4100. For additional information concerning the technical documentation, contact Data User Services Division, Data Access and Use Staff, Bureau of the Census, Washington, DC 20233. Telephone: (301) 763-2074.

# Chapter 12

# Impact of Federal Interventions on Private Mortgage Credit in Rural Areas

*Harold O. Wilson*
*James H. Carr*

Federal interventions in the nation's housing finance system have produced mixed results when it comes to the availability of mortgage credit in rural areas. On the one hand, they have brought mortgage credit directly to rural areas through various direct and insured loan programs, and they have brought greater efficiency to private mortgage markets through standardization in mortgage instruments and underwriting, stabilization of mortgage lending, and encouragement of a secondary market. On the other hand, the supply of affordable credit has not been adequate to meet the need for low- and very low-income housing.

Moreover, the facilitative role played by the federal government clearly has benefited metropolitan areas more than nonmetropolitan areas (also referred to as "metro" and "nonmetro" areas). Federal Housing Administration (FHA)-insured loans, for example, are concentrated in metro areas, as are loan purchases by government-sponsored enterprises (GSEs) operating in the secondary mortgage market. Until the addition of a guaranteed housing loan program operated by the Rural Housing Service (RHS) of the U.S. Department of Agriculture (USDA), federal loan programs earmarked

137

for nonmetro areas largely bypassed the private credit delivery system. Compared with the nation's cities and suburbs, relatively little has been done to leverage private mortgage credit specifically targeted to rural communities.

The Economic Research Service (ERS) of USDA has categorized federal interventions as those that address efficiency issues in rural financial markets and those that deal with equity issues (Sullivan, 1990). This characterization offers a helpful framework for discussing the effects of federal interventions in rural markets. Interventions that improve efficiencies generally are driven by business and financial concerns and designed to increase lender competition, lower transaction costs, or improve access to information. Such interventions often are self-sustaining and include participation of secondary markets, federally capitalized loan or investment funds, technical assistance for borrowers and lenders, bank deregulation, and loan guarantees.

Equity in the allocation of credit is prompted by social concerns for fairness and is addressed by interventions promoting a more equitable allocation of housing resources to support economic stability in rural communities. They include grants and below market rate loans for public infrastructure, direct loan programs to support low- and moderate-income housing development, and grants to support rural housing delivery networks (ERS, 1995). These interventions usually require ongoing targeted government subsidy.

Whether priority in national policy decisions is given to equity or efficiency concerns is driven by budget considerations as well as ideological differences over the proper role of the federal government. In addition, it is influenced by perceptions of the degree to which rural credit needs are being adequately met by nonfederal sources. Even though equity interventions were pursued aggressively from the 1960s to the late 1970s, national housing policy has increasingly moved from retail loan programs dependent on direct government financing and subsidy to programs that rely on the private sector and credit enhancement.

In this chapter, we describe federal interventions that have contributed to greater efficiencies in the allocation of private mortgage credit in rural areas. (For a more detailed discussion of rural credit needs and federal subsidized loan and grant programs, see this book's chapters by Collings [Chapter 9] and Strauss [Chapter 11].) It is argued that the current shift toward efficiency interventions, along with other trends affecting the mortgage finance industry, will lead to an increased flow of credit to rural communities. However, the special nature of rural finance markets, including remoteness of many rural markets, deep and pervasive poverty, the character of low-income housing needs, and the persistent and complex problems of some special population groups, will continue to require direct federal subsidized loans and grants to promote an equitable allocation of mortgage credit.

## The Special Nature of Rural Finance Markets

Rural areas present a special set of problems for private financial institutions. Containing nearly 53 million people, rural settlements differ among themselves as much as they differ from central cities (U.S. Bureau of the Census & U.S. Department of Housing and Urban Development [HUD], 1995). The history, culture, politics, and economics that make up small-town life in Visalia, California, for example, are as different from those of Whitesburg, Kentucky, or Indianola, Mississippi, as they are from those of Newark, New Jersey.

According to Ken Deavers, an economist with ERS, remoteness, low population density, and economic dependence on one industry are three important attributes that bind these diverse communities into what we identify as rural America (Deavers, 1991). To these attributes should be added a predominance of low-income families, special population groups, and poor housing opportunities. All of these have a significant impact on the function of financial markets in rural communities.

Remoteness gets at the heart of the capacity problem in rural America. Rural areas often are not only separated by distance but also disconnected from institutions and resources that urban areas take for granted such as information networks and technical support systems. This isolation fosters a lack of capacity by lending institutions to effectively evaluate risk and undertake complex transactions.

Remoteness also encourages a lack of competition, which can result in higher costs for mortgage credit compared with urban areas. For example, according to the American Housing Survey (AHS), the median interest rate for mortgage loans in nonmetro areas in 1991 was 9.4%, compared with 9.1% in central cities and 9.0% in suburbs. The median loan term in nonmetro areas was 23 years, compared with 30 years in central cities and 29 years in suburbs. Limited availability of long-term, fixed-rate funds, and poor access to the secondary market, result in either 3- to 5-year balloons or adjustable-rate mortgages being the standard mortgage product. Neither is generally acceptable to rural families.

Low population density is associated with the difficulty of pooling loans and achieving economies of scale in lending activity, which can result in higher transaction costs. It also is associated with appraisal issues. Limited sales and lack of comparables in sparsely settled areas create problems for valuation because appraisals generally are based on urban-derived standards that do not fit rural communities. In some communities, appraisals of new construction units do not approach the cost of the houses, thereby requiring gap financing that usually is not available.

Specialization in one industry leaves rural economies and borrowers uncomfortably dependent on a single source of income that is acutely vulnerable to industry collapse, corporate downsizing, restructuring, and/or relocation. This clearly creates concerns about risk on the part of lending institutions, which can further contribute to already limited rural credit availability.

A predominance of low-income rural families offers special challenges that cannot be met by conventional mortgage markets alone. Throughout this century, a disproportionate share of the nation's poor has resided in nonmetro areas. In 1995, the nonmetro poverty rate was 15.6%, which was 2.2% above the metro poverty rate. Even though the rural poor tend to be employed, jobs often are seasonal or in low-paying service industries (U.S. Bureau of the Census & HUD, 1995). Lack of down-payment resources, poor employment opportunities, and low incomes make it extremely difficult for these families to qualify for conventional loans because their credit profiles do not meet standard underwriting guidelines.

Special population groups present similar problems for conventional rural mortgage markets. These groups include farmworkers, residents of *colonias* located along the U.S.-Mexico border, and Native Americans. A variety of conditions, including high mobility, extreme poverty, seasonal employment, land trust issues, title problems, and limited down-payment ability, often keep these families from qualifying for conventional mortgages.

Finally, poor housing opportunities for low- and moderate-income families, reflected in a high incidence of substandard housing and limited choice, also make it difficult for private lenders to serve rural markets. Nonmetro housing units, for example, are nearly three times as likely as metro housing units to fail HUD standards of housing adequacy. For this reason, a higher share of housing in nonmetro areas might be assumed to be unmortgagable on the basis of physical inadequacy (Urban Institute, 1990, p. 13).

# Federal Interventions and Rural Markets

Financial institutions serving rural markets are regulated, supported, and complemented by a number of federal interventions focused on improving market efficiencies and credit allocation. Major interventions discussed in this chapter are represented by programs of the Federal Housing Administration (FHA), government-sponsored entities (GSEs), the Veterans Administration (VA), the Federal Home Loan Bank (FHLB) system, and the Farm Credit System (FCS). (RHS has provided major interventions in rural housing markets through its Sections 502 and 515 programs and has served special population groups through other programs. As indicated earlier, these are not described here but are discussed in detail in Chapters 9 and 11 of this book.)

### Federal Housing Administration and Veterans Administration Programs

The FHA and VA programs focus entirely on improving efficiencies in private mortgage markets. Measured in terms of volume, these programs have contributed more to mortgage availability in urban areas than to low- and moderate-income families in rural markets. According to the 1995 AHS, the suburban share of FHA-insured, single-family mortgages was 54%, compared with 35% in central cities and 11% in nonmetro areas. Over the years, the FHA market share has declined nationwide, primarily because private mortgage insurers began to provide services to upper-income households, the cost of FHA insurance increased, and terms on conventional loans became less restrictive and more competitive with FHA-insured sources (Urban Institute, 1990, p. 7).

Perhaps the most significant impact FHA has had on rural mortgage markets comes from its development of the fully amortized, fixed-interest, level-payment mortgage and its promotion of longer term mortgages and higher loan-to-value ratios. In addition, FHA helped to standardize appraisals and assisted in the provision of information on credit risk (Congressional

Budget Office, 1993b, p. 18). Under the Section 203(b) program, for example, FHA encourages lenders to make higher risk loans (lower down payments and longer terms) by assuming a portion of the risk and to test new
mortgage finance techniques in response to changing mortgage conditions
(National Association of Realtors, 1990). Benefits are extended to rural areas
under a special program, Section 203(i), although urban-based underwriting
criteria have made the program impractical and unusable.

The VA loan guarantee is an entitlement program. Even though some
fees are associated with a loan, no down payment is required and no
restrictions are placed on the loan amount so long as the borrower can cover
the amount over the guarantee limit (National Association of Realtors, 1990).
In 1950, a supplemental direct loan was established for rural areas determined to lack credit. It was terminated in 1981, following the success of a
program to refer rural borrowers with loans approved for direct financing to
private lending sources (Urban Institute, 1990, p. 5). The program, however,
has a dwindling market share and currently little impact on rural mortgage
markets.

## Government-Sponsored Enterprises

With the exception of the FHLB system (discussed in a later section),
GSEs are secondary mortgage market entities that promote efficiencies in
the flow of credit to housing markets by making mortgage loans more liquid
investments, increasing the availability of mortgage credit from capital
markets, and helping to standardize the finance industry. GSEs include
Fannie Mae (formerly the Federal National Mortgage Association), the
Government National Mortgage Association (Ginnie Mae), and Freddie Mac
(formerly the Federal Home Loan Mortgage Corporation).

GSEs do not originate mortgage loans or offer subsidies or grants to
distort the flow of credit. Instead, they purchase or securitize mortgages
originated by others, thereby allowing primary lenders to accelerate recovery
of their loans for additional lending. Both Fannie Mae and Freddie Mac do
support corporate foundations that provide grants to assist nonprofit organizations, but these grants are not directly related to the corporations' mortgage
business.

Fannie Mae was created in 1938 as a federal financial institution to
purchase and hold mortgages as an investment. In 1968, it was made a private

corporation, and its responsibility to purchase government-subsidized mortgages was transferred to the newly created Ginnie Mae. Fannie Mae remains a major market for government-insured or -guaranteed loans, but the overwhelming majority of its total business is in conventional mortgages (Fannie Mae, 1994a). Mortgage-backed securities now represent a substantial part of its business and have served two important functions in making the housing finance industry more efficient (Fannie Mae, 1994a). Ownership of the mortgages is made easier because they are now represented by instruments with characteristics comparable to other securities regularly traded on the finance markets, and they have opened a substantial investment resource through access to the capital markets.

Fannie Mae has undertaken a number of initiatives to stimulate credit for low- and moderate-income families and underserved areas. In 1987, the agency established an Office of Low- and Moderate-Income Housing and began to tailor specific loan products and change underwriting policies to increase mortgage lending to low- and moderate-income families. In 1991, the National Housing Impact Division was established to further emphasize service to low- and moderate-income families and underserved areas. And in 1995, a commitment to invest $1 trillion in mortgage finance for targeted populations by the end of the decade was implemented.

Fannie Mae has a significant rural housing component based on purchases of RHS guaranteed loans. In fact, it has been Fannie Mae's participation in the program that has helped stimulate its growth. In addition, the agency has programs targeted to Native Americans and colonias and has developed specific rural underwriting criteria (Fannie Mae, 1994b).

Ginnie Mae was established as an agency of HUD and guarantees privately issued securities backed by pools of mortgages insured by FHA or VA or guaranteed by RHS. The insurance or guarantee on the mortgages is supplemented with a guarantee on timely payment on securities collateralized by these pools. This pass-through security-linked mortgage originates with capital market investors and helps integrate mortgage markets with bond markets (Congressional Budget Office, 1993b, p. 12). Ginnie Mae participates in rural finance markets to the extent that FHA and VA loans originate in nonmetro communities and through the purchase of RHS guaranteed loans.

Freddie Mac was chartered in 1970 to purchase and sell conventional residential mortgages made by the member institutions of the Federal Home Loan Bank Board (FHLBB), primarily the federally insured savings and loan associations (S&Ls). Following the success of the Ginnie Mae pass-through security, Freddie Mac began issuing similar securities backed by pools of conventional mortgages. The agency now serves the same conventional market as Fannie Mae, although the two differ in terms of the strategies they employ in undertaking their business (Congressional Budget Office, 1993b, p. 22). Like Fannie Mae, Freddie Mac has undertaken an initiative to revitalize neighborhoods. Its Partners in Participation program seeks to create partnerships with nonprofit development organizations and to develop products that respond to low- and moderate-income household needs (Freddie Mac, 1994).

Both Fannie Mae and Freddie Mac have an interest in expanding their low-income and rural markets, prompted in part by affordable housing goals and underserved areas legislation in the Federal Housing Enterprises Financial Safety and Soundness Act of 1992. Under the act, HUD is directed to set and develop regulations to implement affordable housing goals for Fannie Mae and Freddie Mac to better meet the credit needs of low-income families in central cities, rural areas, and other underserved areas. Within nonmetro areas, HUD has identified underserved counties, and loans purchased by the GSEs in these counties will score against the goal.

### Federal Home Loan Bank System

The FHLB system was created in 1932 to provide advances (low-cost loans) to home mortgage lenders to reduce the effects of unexpected withdrawals and stabilize mortgage lending in local areas. Today, the FHLB system continues that mission through a network of 12 independent regional banks. Because of the system's status as a GSE and its low risk, it can raise funds less expensively than can other institutions (Congressional Budget Office, 1993a). As a result, FHLB is a major source of long-term, fixed-rate funds for rural mortgage lenders, and many small banks are joining the system to access the funds. Unfortunately, the funds are finite and loaned out fairly quickly.

Provision of credit for low-income housing is not a principal activity of the system, but two specific programs have been developed to reallocate credit to achieve targeted housing goals. The Affordable Housing Program (AHP) provides subsidies to institutions that finance the rental housing development in which at least 20% of the units are for very low-income families and that promote homeownership for low- and moderate-income families. Members must pass the subsidies through on a dollar-for-dollar present-value basis (Congressional Budget Office, 1993a). The subsidies are based on FHLB's net income; the fixed annual minimum is $100 million. More than $314 million was awarded as of April 1995 and leveraged nearly $4.5 billion (National Rural Housing Coalition, 1995). The program is compatible with the RHS Sections 502 and 515 programs. AHP funds have been used to match Section 502 loans, for example, and to help pay some developer costs for Section 515 projects (Congressional Budget Office, 1993a).

The Community Investment Program is designed to assist home purchasers and renters with family incomes that do not exceed 115% of the area median. Funds also are provided to nonprofit organizations involved in commercial and economic development for low- and moderate-income families. However, between 1990 and 1992, nearly all the funds were used for housing. Unfortunately, it is not known what percentage was used in non-metro areas (Congressional Budget Office, 1993a).

### Farm Credit System

FCS makes housing loans to farmers and farm cooperatives, as well as rural residents, through a system of 25 banks and 267 associations. Funds are raised by selling bonds on a centralized basis in the national money markets and are passed through district Farm Credit banks to associations for lending to individual farmers. Banks in the system currently are allowed to dedicate up to 15% of their portfolios for loans to rural homebuyers. Most of the FCS housing loans carry variable-rate mortgages to avoid interest rate risk. Eligible borrowers are individuals who reside in rural areas; an eligible rural residence for purchase is a single-family, moderately priced dwelling. Loan limits are set on a region-by-region basis. Eligible areas include communities of 2,500 residents or fewer not associated with urban areas (Milkove, 1990).

Proposals have been made to redirect FCS to allow Farm Credit banks to act as conduits between FCS markets and commercial banks, similar to the operation of FHLBs. Farm Credit banks would become limited-purpose, wholesale lenders for commercial banks and FCS associations. Commercial banks would obtain advances by pledging eligible assets as collateral. Assets could include rural housing loans (American Bankers Association & Independent Bankers Association, 1995).

## Federal Policy Trends and Changes in Rural Finance Markets

Following trends set in urban housing policy, emphasis in federal interventions in rural markets has shifted over the past two decades from direct government provision of affordable credit to increased reliance on the tax code for meeting credit needs for multifamily housing and on private lending institutions for meeting credit needs for single-family housing. A shift from categorical to block grant programs also has accelerated. The success of the HUD-administered HOME program, created in 1990, has renewed calls for a HOME-type block grant consolidation of the remaining HUD and RHS programs.

In the 1990s, the movement away from direct housing supply programs also has been reflected in fund reductions for RHS's single-family (Section 502) and multifamily (Section 515) direct-subsidized loan programs, while its single-family guaranteed loan program grew in popularity and lending activity. Use of the tax code to stimulate private capital in affordable housing, however, came under increasing attack. Even though the Low-Income Housing Tax Credit (LIHTC) program has supported most of the low-income multifamily units produced in metro and nonmetro areas since its inception, congressional opponents in 1995 labeled it a form of "corporate welfare" and nearly prevailed in efforts to allow statutory authority for the program to "sunset" (or lapse). Future challenges to LIHTC can be expected as Congress looks for ways in which to reform the tax code and meet balanced budget goals.

Current sentiment among policymakers appears to be that rural housing credit needs can be addressed by improving the efficiency of financial markets without sustaining equity interventions that allocate affordable credit. This sentiment, which is likely to dictate the nature of federal interventions in rural mortgage markets for the foreseeable future, is consistent with a general public sentiment against direct federal subsidies. It also is encouraged by the view of most rural economists that the credit needs of rural America are being adequately met, resulting from four important changes: deregulation of financial institutions, increased rural participation in the secondary mortgage market, new technologies in the mortgage finance industry, and growth in the role of nonprofit financial intermediaries in rural areas.

## Deregulation of the Banking System

Rural housing markets are served by a variety of private lending institutions—commercial banks, S&Ls, mutual savings banks, credit unions, finance companies, mortgage companies, pension funds, and insurance companies (see Strauss's chapter in this book [Chapter 11]). Commercial banks are, by far, the dominant providers of credit in most rural financial markets and for most financial services. In 1989, they originated 65.7% of the loans made by depository institutions in nonmetro areas. S&Ls originated 26.5%, and credit unions originated 7.8% (ERS, 1991, p. 5).

Geographic deregulation of the banking industry has the potential to positively affect rural mortgage markets by increasing the number of local bank offices, encouraging development of bank networks with greater technical capacity and access to capital, and providing greater economies of scale to more efficiently access secondary markets. Indeed, according to ERS, the number of banking firms serving the typical rural county has been on the rise (although not as fast as in urban counties). In the past few years, the typical rural county gained one bank firm and is now served by the offices of five bank firms. The number of counties served by two or fewer bank firms was 20% in 1994, down from 30% in 1986.

With deregulation, consolidation of the banking industry is likely to increase. One result is that more rural counties are now served by banks that are not locally based, and a growing number are served exclusively by nonlocal banks (ERS, 1995). There is speculation that nonlocal banks are primarily using deposits to invest in financial assets that do not benefit rural people. Some argue that smaller, locally controlled banks have better ties to the community and will respond better to local needs. It also might be argued that large financial institutions are better able to offer a variety of services and spread risk over a broader geographic area. Bank holding companies, for example, may raise funds by issuing certain forms of securities, such as commercial paper, and use the proceeds to purchase certificates of deposit from their affiliates (Milkove & Sullivan, 1990, p. 8). Because in the past banks have not had to report mortgage activity in nonmetro areas, it is difficult to judge where credit is being allocated.

Although antitrust laws prevent consolidation within local banking markets, the regulations of the Community Reinvestment Act (CRA) take on an increasingly important role in discouraging large branching banks from siphoning funds out of rural counties or otherwise taking an impersonal interest in their local markets (ERS, 1995). CRA was established in 1977 to discourage the practice of "redlining"—receiving deposits from a particular geographic area but excluding the area from receiving loan services. It applies to federally insured financial institutions that include commercial banks, savings banks, and S&Ls.

CRA regulations implemented in 1996 strengthened the act overall but weakened performance reviews of the small institutions that serve many rural communities. Banks and thrifts are evaluated based on performance in meeting a lending test, service test, and investment test. However, the regulations distinguish between large and small institutions by offering a streamlined examination process for independent banks and thrifts with assets under $250 million or banks and thrifts with assets under $250 million that are members of a holding company with total assets under $1 billion. In addition, small banks are not subject to any additional data collection and reporting requirements unless they opt out of the streamlined assessment procedures. This will allow more rural lenders to fall into the "small" category (Federal Reserve System, 1995).

Experience in the RHS guaranteed program indicates that CRA requirements are helping to expand use of the program because these loans help banks meet the requirements of the act. According to RHS, small banks that normally made only a few mortgage loans to their best customers at 60% to 70% loan-to-value, and for 10- or 15-year terms at best, are now making mortgage loans through the guaranteed program. In addition, large banks have established corresponding relationships with small banks to create origination networks for loans to be guaranteed through the program. Corresponding banks originate loans, and mortgage companies and large banks underwrite and aggregate them for sale to the secondary market.[1]

## Increased Rural Participation in the
## Secondary Mortgage Market

A second important change in rural mortgage markets is expanded activity by the secondary markets, led by participation in the growth of the RHS guaranteed program and prompted by low-income and underserved areas' goals. Participation in and promotion of the guaranteed program by Fannie Mae has been the single most important element in its success. Rural lending institutions had little incentive to participate until Fannie Mae created a market for the investments and began to promote the safety and soundness of the program to its customers.

In addition, Fannie Mae has made a major contribution to rural mortgage markets by establishing underwriting standards that take special rural characteristics into consideration. Problems associated with establishing value in areas with sparse populations and concerns regarding access roads, outbuildings, large lots, and others were addressed. Rural lenders have indicated that the standards are satisfactory in meeting the problems presented by rural areas, but underwriters do not yet believe they are acceptable. As a result, they have not been adequately applied.

Capacity problems in rural areas require that secondary mortgage market entities develop special outreach initiatives to serve these underserved markets. Fannie Mae, for example, has an outreach initiative to the colonias that should be repeated in other rural communities if the low-income goals for underserved areas are to be met. A significant element of Fannie Mae's "Showing America a New Way Home" initiative is the establishment of 25 outreach offices to develop comprehensive investment strategies to address local housing needs. As a result of outreach by the secondary market, RHS's guaranteed program grew from a loan volume of $38.4 million in fiscal year 1991 to nearly $1.0 billion in fiscal year 1996.

The underserved areas' goals might encourage Fannie Mae, as well as Freddie Mac, to enlist private lenders to serve low-income rural markets beyond the guaranteed program. In addition to the low-income products already in use, both corporations will find it necessary to develop specific rural outreach networks to encourage lenders to expand their markets to low-income families (and to assist lenders in the process). It is unrealistic to assume, however, that increased participation by Fannie Mae and Freddie Mac will dramatically affect the availability of affordable credit for low-income families. Grants and subsidized loans, such as those of RHS, are required to serve those markets.

## The Impacts of New Technology

The third change affecting the mortgage finance industry results from rapid technological innovation. Increased efficiencies produced by electronic information networks, loan originations and sales, and automated underwriting enhance the capacity of private lending institutions to serve rural markets. Technology has played a significant role in development of the increased sophistication of the secondary mortgage market in the 1990s. In primary mortgage markets, automation began early in the decade, with information systems offering rate sheets that gave lenders, through real estate agents, a way in which to "advertise" their available loan products and match these products with borrowers' finances. Systems also developed the capacity to offer loan processing and on-line mortgage commitment, even though they generally stopped short of having delegates, such as real estate agents, originate loans. A few systems did evolve that included computerized application processing systems with real estate firms as loan originators.

Entry of nontraditional participants in the real estate industry has brought new methods to enhance customer service and support loan production. Today, interdependence of functions within computerized home mortgage origination systems is pushing companies toward "vertical integration"—combining all services under one roof. Technology can provide instant affordability analysis, comparison of loan products, loan status tracking, and electronic transmission of mortgage applications and related documents (National Association of Realtors, 1990).

Where the industry will go offers fertile ground for speculation. Will it be possible one day, for example, to access a mortgage market network via the Internet and shop for a loan? A rural family in its home or apartment, at the office of a real estate agent, at a nonprofit organization, or perhaps at the post office or library would access an interactive mortgage market network. The shopper would enter the general property location desired or even a specific property, along with personal financial data, to access a list of lenders and their mortgage products including rates, terms, and conditions matched to the shopper's borrowing capability. Products would include conventional, FHA, VA, GSE, RHS, and other products. An application could then be filed electronically along with a credit request. Response, including initial approval, would be received electronically.

The technology to develop such a mortgage market network currently is available, and its development probably is not too far in the future. At the present time, anecdotal information indicates that current technology, combined with networks established by the banking system as a result of deregulation and access to secondary markets, will ameliorate the efficiency problems created by low population density and remoteness that characterize many rural areas. Inasmuch as large banking institutions can spread risk over a larger geographic area and greater number of products, deregulation also can assist those localities dependent on one industry.

## The Growing Role of Nonprofit Intermediaries

Increased reliance on improved efficiencies in the private mortgage industry to serve low-income rural markets will require greater participation by nonprofit intermediaries. Since the late 1960s, rural nonprofit organizations have played an important role in helping very low-, low-, and moderate-income rural families access mortgage markets including the subsidized loan programs of RHS and other federal subsidy programs. By acting as sponsors for housing developments, administering assistance programs, and providing program information, training, application assistance, and housing development services, these organizations have proven to be effective intermediaries between private and public housing resources and low-income families.

As subsidies diminish and increased reliance is placed on private lenders to serve rural low-income housing needs, nonprofit intermediaries will play an increasingly important role in building public/private partnerships to effectively leverage available subsidies and private mortgage resources. For example, several national, regional, and local rural nonprofit organizations have established lending programs for predevelopment costs, rehabilitation, and new construction activities closely targeted to low-income families that often are used to leverage additional funds.

Technical assistance provided by the groups includes financial counseling, mortgage counseling, home evaluation, assistance with evaluating repair or construction needs, securing contractors, and negotiating contracts. In general, these nonprofit lenders tend to offer a full range of services to ensure that quality, affordable construction takes place and that the loans are repaid.

Nonprofit lenders take on increasing importance as private financial institutions stretch to serve low-income markets and as new technology and bank deregulation make lending more impersonal. Nonprofit groups may intervene, for example, if automated underwriting appears to be unfair to low-income borrowers in their area of service. Nonprofit organizations also leverage private funds with their own resources and serve as intermediaries as public funders and lenders attempt to leverage private resources for targeted markets. Fannie Mae, Freddie Mac, RHS, and the FHLB system

have recognized the importance of these organizations by making partnerships with nonprofit community groups one of the keys to their affordable
housing strategies.

## Conclusion

It always is difficult to speculate about future federal housing policy because
unexpected national events can cause public policy to change rapidly. However, trends in current low-income rural housing policy, coupled with developments in rural mortgage markets, indicate that rural housing policy un-
doubtedly will emphasize interventions that stress efficiencies rather than
equity.

This approach will expand mortgage opportunities for some low- and
moderate-income families in single-family markets and will benefit rural
areas in general. However, greater efficiencies in rural financial markets will
not benefit all low-income families and will miss most very low-income
households. Efficiencies also will do little to assist special population groups
that require special interventions. In addition, it is unrealistic to expect
private institutions that carry private sector restraints to shoulder this responsibility alone.

The unique characteristics that bind rural areas into what we call rural
America—remoteness, low population density, and specialization in one
industry—require federal interventions that promote efficiencies in rural
financial markets to ensure an adequate flow of mortgage credit. The predominance of low-income families, special population groups, and limited
housing opportunities require federal interventions that also address equity
concerns in the allocation of affordable mortgage credit.

The question becomes, then, how low- and very low-income families
and special rural populations are to benefit from the present shift in federal
rural housing policy and still maintain the equities of RHS direct interventions. The challenge for policymakers, the private finance industry, and rural
housing practitioners is to develop subsidy mechanisms that take advantage
of private financial institutions and still provide direct interventions in rural
markets that offer both efficient and equitable distribution of affordable
credit for low- and very low-income families. In short, mechanisms to add
subsidies to the guaranteed loan program will be required, as will mecha-

nisms to leverage the remaining direct loan funds via participation with private lenders.

# Note

1. Interview with Rural Housing and Community Development Service officials, Washington, D.C., June 1995.

# Part IV

Creative Solutions

# Chapter 13

# Mutual Self-Help Housing

*Peter N. Carey*

Mention the idea of self-help housing, and it is likely to get a warm, if sometimes skeptical, response. Self-help housing carries with it all the positive connotations of people solving their own problems, working for their goals, and pulling themselves up by their own bootstraps. And, in truth, one cannot help being moved by the vision of a husband and wife working side by side with neighbors to create a better home for themselves and their children. Even the naysayer is apt to say, "It's a lot better than those programs where they just give the house for free." This is precisely why investing in the self-help concept makes so much sense. Low-income families join together in a structured relationship to share in the work of building each other's homes. In so doing, they reduce the cost of the homes, learn skills, and develop neighbor-to-neighbor relationships, which often live far beyond the period of construction.

Organized self-help housing is predicated on four basic beliefs. First is that there is a general social benefit to ensuring decent housing for all. This objective is stated in the Housing Act of 1949 and is the policy basis for federal housing programs. The second is that there is an inherent value in involving low-income people in the solutions to their own problems and a corollary value in the success enjoyed by those who, through building their own homes, achieve a goal met by few. The third is that when given the opportunity, many low-income people can and will work to improve their lives. The final belief is in the power of mutuality—that working together in *mutual* effort to solve *mutual* needs engenders a sense of community and interdependence.

Self-help is popularly seen as an American tradition harking back to the pioneers who set up housekeeping in a new land. It grew out of the "can do" attitudes of these newcomers who, with limited financial resources coupled with determination and the help of neighbors, were able to construct the barns, homes, churches, and schools that became the fabric of America's rural communities. Faced with needs that exceeded the means to meet them, community members would join hands to build together what rarely could be accomplished alone.

In those earlier days, when land was easy to come by and materials were readily at hand, building a home was simpler. Today, we live in more complex times, and even sophisticated homebuilders are faced with a range of challenges from financing to building codes, from environmental issues to land use policy. For those who cannot afford to hire experts for advice, the challenges are overwhelming. Therefore, mutual self-help housing programs, although still retaining a style reminiscent of pioneer barn raisings, provide the organizational structure that allows low-income families to build the homes they so desperately need. This includes the capital, training and supervision, coordination, accounting, and myriad of other technical skills necessary to any successful housing development effort. Today, rural self-help housing programs exist in more than 30 states. They may be stand-alone organizations, or they may be part of social service organizations, nonprofit housing development corporations, or local governments.

With a self-help organization to handle the complexities of development, the way is clear for families to invest their labor—the one resource they have in abundance—as they join with others to build their homes. Over the past 30 years, more than 25,000 of the nation's rural families have participated successfully in mutual self-help homebuilding programs—proof that with the right ingredients, the model works (Housing Assistance Council, 1997a). Organized in construction groups of anywhere from 4 to 12 families, their labor becomes the down payments on the purchases of their homes and on what can become lifelong investments in the lives of their neighbors and the building of their communities. Self-help housing is neither simple nor a panacea for solving the housing ills in rural America. However, it has proven to be a productive and practical avenue to homeownership for families of limited means that might not otherwise be able to achieve this goal.

For more than a quarter of a century, the common denominator in rural self-help housing has been operational funding through the U.S. Department of Agriculture's (USDA) rural housing program, now administered by the Rural Housing Service (RHS). Known as the Section 523 program, it provides grant funds to hire the staff necessary to make mutual self-help housing a reality. When coupled with home loans available through USDA's Section 502 lending program, the two programs offer a financing package that has been at the core of rural self-help housing efforts. It is a model that is replicable in many parts of rural America, raising the possibility that with adequate resources, mutual self-help can become an increasingly important part of this nation's housing effort. What once was seen as a risky experiment has, in the 1990s, witnessed a dramatic new growth in the number of rural self-help housing organizations.

## Beginnings

The roots of organized self-help housing in the United States can be traced to the 1930s and the American Friends Service Committee (AFSC), the Quaker service organization (Margolis, 1967). Since 1922, AFSC had been studying the problems of coal miners throughout Appalachia. When the Depression hit, the demand for coal plummeted, and the lives of coal miners became desperate. With corporate gifts from U.S. Steel and others, AFSC developed plans for a self-help housing project called PennCraft. The concept was broad and included a plan for developing increased self-sufficiency for the 65 families that would live there. AFSC provided the financing, making 20- to 30-year loans at 2% interest.

Other parallel efforts began to emerge around the same time. In 1938, St. Francis Xavier University in Nova Scotia sponsored a self-help housing project for coal miner families (Margolis, 1967). In the decades that followed, 2,000 self-help homes were built. There also were demonstration programs sponsored by AFSC and others in Philadelphia, Indianapolis, and Puerto Rico.

These were largely isolated experiments. The roots of organized mutual self-help homebuilding as a truly national program can be traced to the poor community of Goshen in the heart of California's San Joaquin Valley. Long known as the richest agricultural area in the nation, the valley also was well known as a place of overwhelming poverty in the midst of plenty. Although the agricultural strength of the area produced work for those who tended its crops, it did not produce for them either wealth or opportunity. The housing available to farmworkers in 1960 was part of a picture that had changed little from Edward R. Murrow's portrayal in the *Harvest of Shame* (Murrow, 1960). Those who labored in the fields returned home each day to face the grim reality of their poverty. Their homes, substandard shacks lacking plumbing and heat, often were located in rural enclaves without the basic public services, such as sewer and water, that most Americans took for granted.

As part of its commitment to social services and justice, AFSC had established an office in Visalia, near Goshen. The mission of AFSC staff was to find ways in which to improve the lives of farmworkers. AFSC staff member Bard McAllister worked side by side with farmworkers and routinely queried them as to their dreams and hopes. Universally, the answer was that they wanted decent homes for themselves and their children. From this, the dream began of a concept in which farmworker families would band together to construct each other's homes. What they lacked in financial resources, they could easily offset with hard work and determination.

In 1961, recognizing that mortgage financing was a critical building block of any rural housing program, AFSC staff worked to encourage legislation that would make it possible for RHS to make housing loans to eligible rural residents other than farmers (Housing Assistance Council, 1997a). Prior to that, only farmers could apply for Section 502 housing loans. After the law changed, farmworkers and other low- and moderate-income rural residents also were eligible. In another step forward, RHS agreed to link the Section 502 program with the self-help concept, allowing self-help families to earn credit for the "sweat equity" earned by providing the labor to build their own homes. It took some convincing, and at first RHS would finance only the houses, not the land on which they were built. For the first homes, AFSC carried second loans on the land. But, ultimately, the model was proven, removing one of the basic obstacles to widespread self-help housing. Three decades later, the Section 502 direct lending program remains the cornerstone of rural self-help financing.

The success of the first 20 families in Goshen, working in three separate groups, led to the decision on the part of AFSC to find a means of funding the organizational infrastructure that would coordinate and expand the program. With passage of the Economic Opportunity Act of 1964, which launched the "War on Poverty," a potential funding resource emerged. In February 1965, Self-Help Enterprises (SHE) was incorporated as the first rural self-help housing organization in the nation, and just weeks later, the U.S. Office of Economy Opportunity (OEO) awarded a self-help housing grant to the newly formed organization. This was to be the first of many such grants made under OEO's Title III-B migrant program to rural organizations throughout the country.[1] The concept of building one's own home was not new, nor was the notion of shared labor. What was new was a national commitment to support participants in the process and to recognize that their labor could create capital in that basic American investment—a home.

## The Mutual Self-Help Model

There is no single model for the mutual self-help housing project. Program design varies from region to region and must take into account everything from local housing standards, to land use regulations, to financing availability, to the nature of the local workforce. However, in some fashion, the program design must address each of the following components: availability of building sites, access to affordable construction and mortgage financing, identification of qualified participants, skilled construction supervision, and the organizational capacity to coordinate the effort.

### Building Sites

Mutual labor exchange is much more easily managed in a situation where the homes are built on contiguous sites. This allows for more effective support by the on-site construction supervisor and also makes it easier to overcome the natural tendency of families to concentrate on their own homes rather than on each other's homes. However, there are areas in which large lots, the configuration of land uses, and geography make contiguous groups all but impossible. These areas place an extra burden on the program.

In other areas, subdivision-type development might be possible; however, there might be difficulty in obtaining adequate lots, and the organization might not have the financial or technical capability to engage in its own

land development activities. Some organizations do develop their own sub-
divisions, and this creates its own set of technical demands on the skills of
the organization and also is quite capital intensive. Whatever course is
chosen, buildable lots must be available at the time construction is to start,
and title must be vested either with the homebuilder or with the sponsoring
organization.

### Financing

The ideal lending program for mutual self-help continues to be USDA's
Section 502 program because it is a convertible loan that closes prior to the
beginning of the construction. Thus, the loan can pay for the lot, materials,
and any subcontractors necessary during the building process. At the completion of construction, the loan converts to a permanent mortgage, and the
family moves in. For low-income borrowers, the program provides an interest subsidy, thereby ensuring affordability. Currently, the amount of subsidy
is pegged to a family's income and is figured on a graduated sliding-scale
basis. For example, a family at 50% of the median income pays the amount
due on a 1% loan or 22% of income, whichever is greater, whereas a family
at 80% of median income pays the amount due on a 6.5% loan or 26% of
income, whichever is greater.[2] Borrowers are requalified annually, and subsidy is decreased as income rises. In the rare case in which income drops,
there is the possibility of increased subsidy to maintain affordability and
protect the investment in the home.

The impediments to use of commercial lending for self-help housing are
significant due to interest rates, lack of subsidy, and the difficulty many
lenders have in assessing the repayment ability of borrowers whose employment might be seasonal. Moreover, commercial lending generally is available only at completion of the home, necessitating interim construction
financing. The interest costs on construction financing are an issue for any
builder or developer. However, in the case of sweat equity homebuilders, the
impact is magnified because of the length of time it takes to complete
construction. Finally, many lenders refuse to treat sweat equity as true equity,
that is, with a value as solid as the cash down payment used by mainstream
homebuyers. As a consequence, the Section 502 program remains the best,
and in most areas it is the only, comprehensive lending program suited to
self-help construction.

**Participants**

Long before construction can start, participants must be identified and organized into a cohesive group. The preconstruction effort usually begins when building lots are available. In some communities, recruiters must actively seek participants. In other areas, there may be waiting lists of families wanting to build. The window of eligibility is rather narrow. Participants must be able to meet the creditworthiness standards of the lender, demonstrate reasonable repayment ability, and have the physical ability to carry out the work.

Due to the fact that mutual self-help housing is geared toward people whose incomes are low or very low, more often than not, there may be a variety of problems in a family's credit report—problems that need to be resolved prior to submitting the loan application. Low-income families typically are unaware of the implications of credit problems. Repayment ability also might be difficult to determine because of income that is unstable or seasonal and, as in many rural areas, based on the economic vicissitudes of a single resource-based industry. Therefore, it is important that both financial counseling and preparation for the responsibilities of homeownership be part of the self-help program.

Perhaps hardest to assess is the ability to carry out the labor of building a house. Although it comes as a surprise to many that the construction process is not technically complex, the work is hard nonetheless and, at times, may seem endless. Families must balance labor requirements with their regular employment—after work, on weekends, on holidays. Each family, in general, must commit to 40 hours of labor per week. Most of that time must be committed by the adult members of the household, although it is not unusual for relatives, friends, and neighbors to assist in the homebuilding and to contribute the hours necessary to meet the family's labor commitment to its group.

**Construction**

Many of the tasks require labor supply beyond that which could be provided by the individual self-help family. In mutual self-help, this labor pool is provided by the group under a shared labor agreement. In more individualized self-help-type programs, that labor may be provided through volunteers or a job training program.

As mentioned previously, the self-help concept hinges on the fact that actual construction of a tract-style home is not technically complex. From a practical point of view, most repetitive skills are easily acquired by reasonably "handy" individuals, and few tasks cannot be completed successfully by self-help homebuilders. In each program, however, some tasks generally are subcontracted out. Which trades are best subcontracted depends on the type of construction, the organization's skills, and local regulations. For instance, in some programs, participants do the wiring; in others, the wiring is contracted out to electricians. There also might be complications with local building departments, or even state law, that require certain aspects of the work to be performed by licensed professionals.

To be successful, the group must be overseen by a qualified construction person. Chances are, the construction supervisor also will need to be something of a specialist at group dynamics. As the work progresses, the on-site supervisor faces a special challenge of working with unskilled workers in the hours that are available outside of their regular employment. Tensions naturally grow out of the dynamics of the group, the pressure to keep the work moving, the interdependence necessary to accomplish the work, and even the new roles that may be defined as men see women doing work traditionally seen as "men's work." As with any group effort, there is ample opportunity for the negative aspects of group dynamics to surface.

In the most successful groups, the labor is mutual. Construction of each home in the group progresses at the same pace, first completing all the foundations, then the framing, then the roofs, and so on. This keeps up momentum, avoids singling out individual participant families, and maintains the collective commitment of the group, which is the key to success. It also allows peer pressure to work effectively to move each family along at the level of participation necessary.

### The Organization

Finally, there is a basic need for the organizational structure to provide the staff and the support system around which a self-help housing effort can occur. Regardless of the size of the program, which may range from 10 to 200 homes per year, the complexities of development must be met. In addition to the organization's role in acquiring building sites and financing for the homes, there is the need to ensure compliance with any number of local regulations, such as building codes and zoning ordinances, and to handle the bills for building materials, subcontractors, insurance, and reporting. Although there are variations around the nation among self-help housing organizations, the model generally involves not only the use of staff with construction skills but also program administration and accounting skills. In the smallest organizations, however, one person may fill all of these roles.

## Challenges

Despite a generation of demonstrated success, the challenges of today are no less daunting. Self-help housing programs face all the complexities inherent in conventional housing development as well as some challenges unique to the programs. At a time when the appropriate role of the federal government in meeting the national goal of decent homes for all is being questioned, rural self-help housing continues to rely on federal support through the programs of USDA, particularly Section 502 mortgage financing and Section 523 technical assistance funding. No parallel financing has been introduced for communities that fall outside of the USDA definition of a rural community. In many areas, self-help programs may work with a variety of other mortgage resources, often through state housing finance agencies, but the fact is that rural self-help housing is uncomfortably dependent on a single source of financing.

In addition, the costs of construction continue to escalate. Development standards are becoming more strenuous, and cash-strapped communities are applying development fees to cover everything from sewers and road construction to libraries and fire service. In many areas, land costs are on the rise. The growing prevalence of significant up-front expenses is resulting in higher interim financing costs. For this reason, a reasonable source of below-market interest rate interim financing is critical. As these other costs increase, labor is decreasing as a portion of the total cost of construction. Consequently, the value of the sweat equity contributed by families and its utility in reducing the overall cost of purchasing homes is less dramatic today than it might have been 10 to 20 years ago.

Increased costs require increasingly complex financing models, both for construction and for mortgages. It is not at all unusual for second- and third-position "soft" loans to be needed to obtain a financing package affordable to low-income, self-help builders. This has become the rule of thumb where Section 502 loans are unavailable, dramatically increasing the complexity of the project and the expertise required by the self-help organizations.

Lastly, NIMBY (not in my back yard) attitudes in some rural areas, deriving from the economic and racial prejudice that often has dogged low-income housing efforts, magnify the difficulty of securing adequate and affordable building sites and might place self-help organizations in an adversarial role with the local communities. Opponents lean on the usual arguments that confront low-income housing development anywhere as well as the myth that self-help construction will reduce the property values of nearby homes. Onerous delays and conditions placed on the issuance of building approvals to placate community opposition add more cost to the self-help homes and stigmatize the future occupants of the housing.

## Conclusion

Unfortunately, self-help housing organizations in the past decade have found little time to reflect, that is, to research and review. So, beyond some early studies, much tends to be anecdotal about the presumed stability of self-help neighborhoods, the success of self-help in building family wealth, and other positive outcomes.[3] There is a need to revisit the concept and collect the data necessary to support what self-help homebuilders know firsthand—that the public investment is reasonable and the public benefit dramatic. Certainly, no one with experience would argue that organized self-help housing is the easiest way in which to produce housing. The issues of organizing groups, managing group labor exchange, and building homes with unskilled labor create complications. However, the results go far beyond the physical and the financial.

Some 30 years ago, the visionaries who founded the first self-help housing program in California debated the mission. Was the goal to produce houses, or was it to produce homes, neighborhoods, and communities? The answer seems simple enough. Children grow up in homes, not houses, and the fabric of the nation is woven from the threads of neighborhood and community. There are abundant examples of what happens when the sole focus is production without the leavening of social principle. Sweat equity builds the house. Mutual self-help builds the community.

Mutual self-help housing broadens the horizons of its participants, helping them to see their own potential and the potential within their neighbors and other members of their communities. Renters who have lost hope of decent places to live because they were at the mercy of negligent landowners find themselves in a position of self-determination. They have created, with their own hands, these places they now call home. Self-help homebuilders know, in ways homebuyers never will know, what it took to make their houses what they are. And it is up to them to determine what types of homes they will be 5 or 10 years down the road.

The sense of self-determination and responsibility extends to the neighborhood as well. Self-helpers have sweat (and usually some blood) invested in the homes of their neighbors. The intensity of the building process has brought them into a working relationship with the people among whom they live. The children have played together while the parents worked. They know each other's strengths, weaknesses, needs, and aspirations. They have seen in concrete ways how their own success is bound up together with the successes of their neighbors. That lesson is not easily forgotten. When the lesson of interdependence is extended into an interest in a participant's larger community, the goal of mutual self-help housing has been fully realized.

## Notes

1. The creation of the Section 523 program in 1970 began a gradual shift from OEO to RHS as the source of funding for the technical assistance program. Establishing a grant program for construction supervision within the jurisdiction of USDA, it was argued, would link technical assistance directly to the RHS Section 502 loan program, the major source of financing for self-help home construction and acquisition in rural areas.

2. As of 1996, RHS reduced subsidy levels and adopted a subsidy calculation calibrated to different income bands. Currently, new borrowers pay the greater of (a) an amount due on a specified reduced interest rate or (b) a specified minimum percentage of adjusted family income. Existing borrowers still pay the greater of (a) the loan payment at 1% interest rate or (b) 20% of adjusted income minus real estate taxes and insurance. The amount of the subsidy paid by RHS is the difference between the amount due on a note rate and the required borrower contribution as calculated here. See 7 CFR § 3550.68(c) and (d) in the *Code of Federal Regulations*.

3. However, a study performed by SHE in 1997, *Self-Help Enterprises and the Mutual Self-Help Home Ownership Experience: 1965-1996*, does support the anecdotal evidence that

self-help homebuilders achieve remarkable stability. The study focused on the ownership history of more than 1,000 California families that built self-help homes between 1965 and 1996. In a state where the median period of homeownership is about 7 years, 78.8% of SHE homebuilders exceeded the median. In fact, more than two thirds of the families that built their homes between 1965 and 1975 still were living in their homes 20 years later.

# Chapter 14

# New Housing Forms for Low-Income Rural Families

*Patricia Harrison*

Today, especially in rural parts of states experiencing rapid growth, producers of affordable housing face serious challenges. Expanding suburbs have driven up land costs, immigration has changed the composition of the rural population, deteriorating local infrastructures and inadequate community resources are stressed to their limits, funding sources for housing are reduced due to political pressures, economic shifts have increased joblessness among the rural poor, and community resistance to the permanent residence of low-paid minority workers and their families is strong. In light of these challenges, new strategies for the creation of affordable housing are being explored including the linkage of housing projects with regional economic development programs, public-private sector partnerships that redefine the role of government, and land trusts.

Housing design, however, is one aspect of the development process that seldom is addressed in efforts to meet the demand for affordable rural housing. If dwellings can be made smaller and more efficient, if sites can be used more effectively by increasing density, if mass-produced dwellings can lower labor costs, and if housing forms can be targeted for specific populations, then an overall reduction in development budgets can be achieved.

169

Moreover, if affordable housing needs can be factored into comprehensive planning for rural development, then the social impact of this new thinking could be significant for community stability and managed growth.

Such an approach, however, also requires critical rethinking about the centrality of single-family home provision in most efforts to address the shelter needs of the rural poor. Evaluation of the range of living circumstances within this population indicates that a variety of housing options, in addition to the single-family home, could be of benefit to families as they become established in a community, obtain regular employment patterns, enroll children in schools, and take advantage of health, citizenship, language training, and other support services (Harrison, 1995). Recognition of a socioeconomic "ladder" in housing, and creation of dwellings responsive to the unique circumstances of each life-cycle step in families' economic development, would offer families the best chance to secure housing that is both appropriate and affordable. And, given increased land and development costs and reduced funding, this might be the only way in which to house a large percentage of the rural poor for whom the single-family home no longer "pencils out."

Ironically, affordable housing providers seem to pose the greatest resistance to development of housing forms that differ from the single-family prototype. In large part, this resistance might be due to their long struggle to achieve this goal. For decades, they have labored hard and successfully to create or restore small and decent homes that reflect the housing norms of the surrounding community and promote community inclusion for low-income families. Within the rural context, this usually means that low-income housing must mirror typical middle-class, single-family housing.

This type of housing stands in stark contrast to the impersonal, barracks-style projects (often associated with public housing) that are easily identified within a small community by their repetitive forms, unfenced and ill-defined open spaces, and large concentrations of low-income families. Although most affordable housing projects today create a positive, integrating appearance in rural communities, the negative image of old-style public housing persists. Some nonprofit developers, although recognizing that the denser and more efficient dwellings of this style of housing might be more economically feasible, have been wary of moving in this direction in the belief they cannot achieve community acceptance.

This chapter proposes new housing forms with the potential to provide decent shelter, reduce project costs, and overcome community opposition.

Although using examples from California, these forms should have relevance in other states where it has become increasingly difficult to produce housing compatible with the income and household needs of low-income rural wage earners. It is argued that it is possible both to design more economical dwellings (e.g., smaller, denser, shared facilities) and, at the same time, to maintain the residential, neighborhood-fitting characteristics of single-family homes.

## New Housing Forms

Several new forms of housing already have emerged in both rural and urban settings—hospices for ill or dying individuals, battered women's shelters, single-parent shared housing, senior housing, congregate housing for the physically or mentally disabled, single-room occupancy housing for the chronically unemployed, co-housing for middle-class families, and transitional housing for formerly homeless individuals, substance abusers, or parolees. These forms have earned acceptance for their appropriateness to community issues and their capacity to meet household needs at various points in the life cycle. In addition to its particular building configuration, each housing type has its own social program and support network that recognizes special circumstances but also works to integrate individuals, to the extent possible, into the community mainstream.

A similar user-focused approach to housing for the rural poor seems quite reasonable within the framework of "special" housing forms. Home-ownership and market-rate rentals are far beyond the financial capacity of most low-income rural families that live at subsistence level, struggling to piece together enough short-term employment to meet basic needs (Harvard University Joint Center for Housing Studies, 1997; "The State of Low-Income Housing," 1997). Existing shortages of rental housing, costly initial rental deposits, rental housing discrimination, and the necessity of moving between regions in search of jobs are among the factors forcing many families to pay exorbitant rates for the privilege of living in substandard conditions.

Framing new housing forms around the specific needs of today's rural poor could be a way in which to house people decently, permit them to contribute to the costs of their homes in proportion to their earnings, and gain greater community acceptance. New housing forms also might be able to capitalize on the cultural norms of immigrant populations for cooperative, denser housing arrangements and shared facilities. Interviews conducted by this researcher with low-income farmworkers indicate that access to clean, safe, affordable shelter is a primary concern for families as a first step in building family wealth (Harrison, 1995). Over time, as families become more financially stable, they hope to move into larger, more substantial housing.

For new housing forms to be successful economically, meet the affordability requirements of low-income families, and be integrated within the context of rural communities, five basic design criteria must be realized:

1. Dwellings must be smaller.
2. Sites must be more densely populated.
3. Construction methods must be "mainstreamed" and new material applications developed.
4. Dwelling design must be carefully fitted to the social and economic means of low-income people.
5. The dwellings and site must render the project a "good neighbor" in terms of their appearance, their scale, and the number of residents on the site.

In the following discussion, five housing forms are proposed that are intended to meet these design criteria: (a) "enclaves" of very small family dwellings; (b) family-oriented, multifamily townhouse and apartment projects; (c) seasonal farmworker hostels for families and single males; (d) dorm-style, communal facilities for families and single men; and (e) roadside rests with short-term camping capacity.

### Enclaves of Very Small Family Dwellings

One alternative to more costly single-family home provision is the creation of small land parcels with enclaves of very small family dwellings within an existing community. Site density would include between 10 and 12 families per acre in two- and three-bedroom, freestanding dwellings. Two-bedroom homes would range from 700 to 800 square feet, and three-bedroom homes would range from 800 to 900 square feet (Figure 14.1). Each dwelling would have its "family domain" in a semiprivate rear yard, a private front porch, and a designated parking shelter, but it would share common front yard landscaping, a small community building, and laundry facilities. Dwellings could be manufactured off-site and set onto permanent founda-

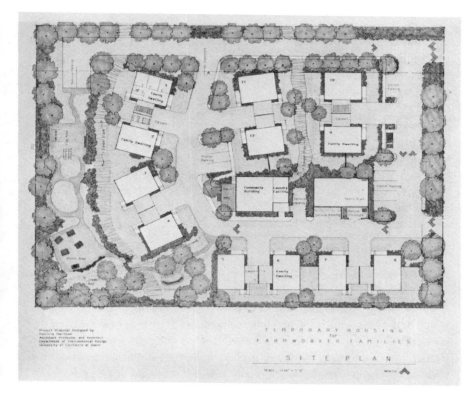

**Figure 14.1.** Site Plan of Very Small Dwelling Enclave

tions to reduce construction costs. Coordinated colors and surface treatment could provide variation and individual identity. Extensive landscaping with trees, lawn, shrubbery, and perimeter fencing would integrate the site and define its limits within the larger neighborhood. On-site recreation areas such as play equipment, a barbecue area, and a basketball hoop would provide focal points for children's and group activities.

Several benefits accrue from the enclave housing type. The relatively small size of an enclave reduces the perception of a "project" with a large, concentrated population of low-income families. In-fill sites within existing neighborhoods become opportunities to integrate low-income families on a scale that is less intimidating to the established community. In more rural settings, enclaves can take advantage of smaller parcels of land and reduce the impact on infrastructure, traffic, and social services. Depending on the economic potential of residents, a lease/option-to-purchase arrangement or cooperative might be substituted for rental options. Careful selection of residents and effective support services would ensure that the enclave was well maintained and safe.

A variation on the enclave theme is a group of semi-attached bungalows, again with common areas and facilities and with a parking garage or carport structure located at a perimeter or site entry point. Urban cooperatives have experimented with this housing format in renovated hotel and industrial buildings with considerable success. Often, such projects have commercial outlets at the street level to provide informal gathering points for residents as well as employment opportunities. In a rural setting, this type of enclave development is uncommon but has the potential to achieve similar success.

### Family-Oriented Multifamily Projects

Two-story townhouse projects have become a cost-effective way in which to provide family housing with some of the amenities of single-family homes. Village Park, a 50-unit complex developed by the Rural California Housing Corporation in Sacramento, California, provides an example of this housing approach. Families have small, private backyards, adjacent covered parking, and garden plots. Common areas include a small community building, a playfield, and tot lots for small children. The project design was conceived around the cultural patterns of Laotian refugee families and the theme of "street face and village core." It attempted to provide a conforming exterior appearance to integrate the housing into the neighborhood, while incorporating interior spaces that would support traditional village activities such as gardening (Figure 14.2). Unfortunately, the design did not account appropriately for typical large family sizes, and parents have complained that their older children must leave the site for play activities.

Crossroad Gardens, a 70-unit, low-income housing project in Sacramento County, California, has drawn heavily on the lessons learned from Village Park. It offers a mixture of apartments and townhouses in which

**Figure 14.2.** Village Park Rear Yards and Communal Gardens

residents have substantial on-site recreational and community spaces for families and children of all ages (Figure 14.3). A child care program for 75 children (including a Head Start program) is housed in a large community building. A half-court basketball area and large playfield are available for all residents in addition to fenced child care play yards. Two tot lots, community gardens, a picnic area, and a soccer field also are included on the site. With a site population of 400 people, the goal is to create a neighborhood within

**Figure 14.3.** Crossroad Gardens Apartments and Townhouses, Showing Expanded Recreational Facilities for Residents

the project that will provide balanced opportunities for family privacy and age-related communal activity. Although dwellings are quite small (a typical three-bedroom apartment is just over 1,000 square feet), attractive and varied unit design, outdoor site amenities, and a well-developed landscaping program evoke the feeling of a residential community with strong middle-class characteristics.

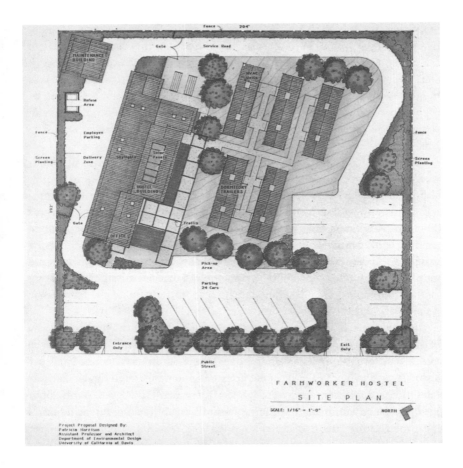

**Figure 14.4.** Farmworker Hostel Site Plan. The plan shows main building for office, food service, and laundry facilities, and portable buildings with dormitories and toilet facilities.

## Seasonal Worker Hostels for Families and Single Males

Short-term housing sites for seasonal agricultural workers have declined seriously in the past three decades. From a grower's perspective, the increased reliance on labor contractors to deliver workers to the site when needed, onerous housing regulations, and liability insurance for on-site employees have been the chief factors in reducing the number of bunk dwellings for workers. Farmworkers' increased mobility has enabled them to live in community settings with rental opportunities and to drive to work sites, but not enough seasonal housing has been developed to replace the loss of grower-operated housing. Competition for an insufficient number of affordable rental dwellings forces many to maintain year-round residence in a "home base" area and to tolerate short-term, seriously substandard housing in other areas.

Although migrant laborers may prefer other housing forms for their primary residence, simple hostels could provide decent, short-term shelter as they engage in seasonal employment away from their home communities. A hostel would be a complex of minimally outfitted dwellings (Figures 14.4 and 14.5). For men, it might include dormitory trailers and a main building with dining hall and laundry services. Family housing might be provided in manufactured units. During periods when migrant workers are not present, the dwellings could be converted to disaster emergency housing for the community or cold weather emergency or transitional shelters for the homeless.

An example of the hostel format is Arvin Village in the town of Arvin, California (Harrison, 1994). Opened in 1992 in an intensively cultivated area with few affordable housing alternatives, this community-operated facility is similar in concept to the seasonal housing centers operated by the California Office of Migrant Services (OMS) for more than 5,000 farm labor families. The dwellings are trailer-style units, each housing either a family or six men. Unlike the financially strapped OMS centers, which are open for 6 months of the year and have subsidized rents of approximately $250 per month, Arvin Village is run year-round and charges $450 per month. In spite of the relatively high rent, a lack of support programs for children, and poor construction and design (with units in rigid rows), the units are regularly occupied during the peak growing season, although occasionally emptying in the winter when workers are unable to find employment.

What Arvin Village demonstrates is the possibility of a "franchise system" of year-round, small rental centers for seasonal farmworker families. Such a system could be instituted with an attractive design format, mass-produced dwellings, and standardized management and maintenance system that could serve the low-income agricultural worker community. With rental

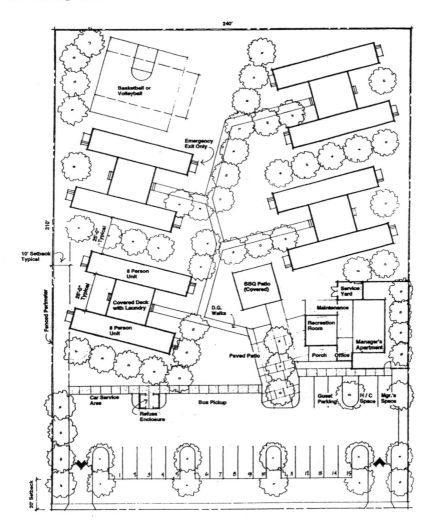

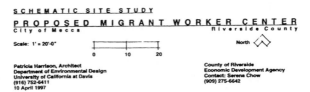

**Figure 14.5.** Site Plan of Seasonal Worker Center for 64 Men

subsidies provided by local growers or housing authorities, and with health and educational support services, the Arvin Village concept could become workable in many areas.

A similar farmworker hostel format focusing on male-only, short-term residence for groups of 40 or fewer men also would be appropriate in agricultural areas with short, labor-intensive periods that do not attract entire families. Again, the design goal would be to house men in residential, rather than institutional-looking, structures on a carefully landscaped and fenced site with on-site parking and laundry facilities. Key to community acceptance is a compatible appearance and a manageable number of residents. Although units would be small and basic in nature, they would offer clean, safe, affordable opportunities for low-income workers.

## Dorm-Style Communal Dwellings

Many immigrant families are familiar with shared facilities where cooking and socializing spaces are open to casual interaction. In California's Central Valley, for example, there are concentrations of Mixtec Indians from Central Mexico living at subsistence level whose housing preferences often are expressed in terms of a site where they might live in a communal group. While some group members are working, other men and women would perform household, cooking, and child care tasks for the group. A housing environment that fosters mutual support (i.e., permits group cooking, dining, and living activities) would help very poor rural families achieve a measure of economic security that they could not achieve alone. This format has been successful in single-parent group homes and family homeless shelters.

Dorm-style communal dwellings could be accommodated in rehabilitated hotels or commercial buildings in rural communities. By centralizing kitchen and bathroom facilities, cost savings could be realized. Providing each family unit with a sitting room and sleeping spaces, and pairing single men in dormitory rooms, would meet basic housing requirements and offer reasonable privacy. Clearly, supervision and structured behavior and personal responsibility issues become more significant in more concentrated living conditions, but common lifestyle and economic circumstances could contribute to a very successful residential environment.

### Roadside Rests With Camping Spaces

In some areas, low-income seasonal workers migrate in family, extended family, or all-male groups. They travel in cars and campers, sleeping outdoors in makeshift encampments or crowding into motels. Migrants have said that it is relatively easy to find meals, but opportunities for daily showers, toilet needs, and laundry services are rare. Similarly, places that welcome short-term labor encampments also are difficult to find, and when found, they often are severely overcrowded.

A system of roadside rests designed around short-term camping, bath, and laundry functions could offer valuable services to families and men on migrant labor routes. Rest facilities would include men's and women's rest rooms, laundry sinks, telephones, bulletin boards with pertinent announcements for farmworkers, and a maintenance and mechanical area (Figure 14.6). Adjacent to the rest would be a parking and camping area with designated spaces, fire pits, clotheslines, a water supply, and a refuse area. Construction materials could be mainly concrete components with other durable, low-maintenance elements. Use of solar water heating and skylights could reduce energy requirements. Although the requirements for plumbed facilities involve considerable cost, fees paid by workers for showers and camping privileges, as well as by hunters, fishermen, and campers in the off-season, might defray some of the expense of regular cleaning, security patrols, and upkeep.

A variation on this concept was tried in Sonoma County, California, by the California Human Development Corporation (CHDC). Seasonal laborers harvest grapes in this area every year, and many find housing unavailable. CHDC rented a campground and installed tents. Kitchens, rest rooms, and showers were installed temporarily. Although the encampment was deemed too makeshift by county officials for year-round use, it was quite successful in the short run. Workers were happy with the security of the arrangement, behavior problems were minimal, and less satisfactory alternatives were avoided.

## Conclusion

Key to development of housing alternatives is general acceptance by affordable housing developers and public agencies that not all rural, low-income

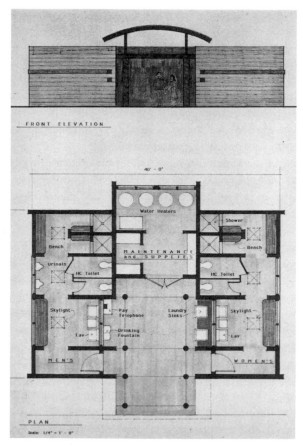

**Figure 14.6.**  Plan and Elevation of Simple Roadside Rest Building. The building was constructed with low-maintenance materials and energy-efficient planning.

people can live in detached, single-family, owner-occupied homes. However, families and individuals at every economic level need homes from which to develop their families, conduct their productive activities, and participate in community affairs. Having a home that is very small, rented, or part of a group complex does not permanently consign a family to a particular housing environment. Just as elderly adults move into senior housing at a certain point, so too can low-income families move on to larger and perhaps more conventional housing forms as their economic means permit. Recognition of the life cycle of residential choices, and understanding that different approaches may be appropriate in different areas, must inform decision making about new housing alternatives.

Although new housing forms proposed in this chapter would be smaller and denser, constructed in an industrial setting and transported to sites, and developed in in-fill sites for better community acceptance, all must meet certain "nonarchitectural" requirements to be successful. First, well-designed management programs must accompany each form. Support programs that foster cooperative and healthy behavior by residents also must be offered where they are needed. The use of shared resources, such as laundry facilities, community rooms, and recreation areas, also must be clearly defined to reduce conflict among residents. Low-income immigrant families might need instruction in everything from household maintenance, to language instruction, to lifestyle norms of American culture. The importance of such support in any housing environment cannot be underestimated.

A second element in the design of new housing forms is provision of a range of spaces for families and individuals. Needs for personal privacy, family interaction, and gender- and age-based activities must be present in discrete zones on a housing site. Clearly demarcated areas for families should be fenced or otherwise marked, and places for individuals to comfortably observe the activities of the housing site without direct participation (e.g., benches, front porches) should be provided. Play areas for children of all ages and places where several families may congregate for a special occasion, and men or women may gather for conversation or recreation, are some of the ways in which residents of small dwellings can find extended space for personal development.

Third, low-income projects need to be designed on a human scale. Although there no doubt will be similarity of dwelling types, personal identity may be given to individual units through differentiation of form, off-setting dwellings from one another, and color and material variation; this can lend a neighborhood quality to the project as a whole. The dwellings must be residential in character rather than barracks style. Elements of single-family homes, such as front doors, formal access points for visitors, and separation of informal family spaces from common areas, must be created.

Finally, housing must have a neighborhood identity that respects the character of surrounding residences. Attention to relative scale of dwellings, attractive colors and materials, and generous landscaping are important to support inclusion in established neighborhoods.

It might seem regressive to back away from the dream of a single-family home for every low-income rural family. However, the realities of current economic conditions and the increasing numbers of very low-income residents make the search for alternatives practical and worthy. New housing alternatives can offer positive results. If housing can be viewed as representative of a level on the socioeconomic ladder, then individual and family progress can be gauged by movement from an extremely basic dwelling to the ultimate goal of a private home. Rural communities can develop more sophisticated understanding of the role of low-income wage earners in the rural economic fabric and awareness that the goals of low-income families for achievement and community participation are similar to their own. If families are accepted as full-fledged community members, then they can have pride in their status at any level and aspire to move ahead as their circumstances improve. They can become more productive and stable if their needs for secure and decent housing can be met at a level appropriate to their means.

# Chapter 15

# Community Land Trusts and Rural Housing

*Kirby White*
*Jill Lemke*
*Michael Lehman*

The community land trust (CLT) model was developed in the 1960s by the Institute for Community Economics (ICE) as a nonprofit vehicle through which geographic communities could gain greater long-term control over the use and allocation of their land, thereby ensuring not only environmentally sound land use but also affordable access for lower income people.[1] Although the model is now associated primarily with affordable housing programs—both urban and rural—it originally was conceived to address a range of land uses. These included housing as well as other, primarily rural uses such as agriculture, production of fuel wood and lumber, and preservation of open space and ecologically sensitive areas.

The CLT differs from other nonprofit housing and community development organizations in its distinctive approach to the ownership of real estate and its governance structure. When a CLT acquires real estate, its intention is to hold the land permanently in trust for the benefit of the community. However, improvements on the land may be owned by individuals or groups within the community, who lease the land from the CLT, typically for a renewable 99-year term. The ground lease gives most of the rights of ownership to the lessee but allows the CLT to retain control over the future use and allocation of the property. The lease normally requires that lessee-homeowners live in the homes for as long as they own them and restricts sales to households with incomes below a stipulated level for prices that are limited by a specified formula. Each CLT designs its own resale formula, but in every case the goal is to strike a reasonable balance between the home-owner's interest in a good return on the investment and the community's interest in seeing that homeownership remains affordable for other lower-income people.

The CLT's governance structure also is intended to strike a balance between potentially competing interests. The CLT is designed as a broad-based membership organization, with membership open both to those who occupy or use CLT land and to others in the local community, who may be neighbors of CLT lessee-homeowners or may wish to become CLT lessee-homeowners. The board of directors is structured so that one third of the directors represent the "resident membership," one third represent the "non-resident membership," and one third represent "the broader public interest." Thus, the interests of current CLT residents are balanced with the interests of potential future residents and others in the community.

As the details of this model were being developed in the 1960s, ICE began working with civil rights leaders in southwest Georgia to establish what would come to be known as the first CLT. The goal of this organization, New Communities Inc., was to provide access to land, housing, and economic opportunities for southern black farmers for whom the lack of such access was a fundamental problem. In 1969, New Communities purchased a 5,700-acre tract of land (for more than $1 million) and developed a plan that included housing, agriculture, and other economic activities. However, the federal support that was hoped for did not materialize (beyond an initial planning grant), and the unsubsidized cost of the land resulted in a debt burden that finally led, after more than a decade, to the failure of this ambitious effort.[2]

Since that time, other rural CLTs have been motivated by the same vision that motivated New Communities—the vision of access to land, housing, and economic opportunities for low-income and disenfranchised people. Most of these organizations, however, have undertaken more modest programs involving acquisition of smaller parcels on a scattered site basis, with somewhat less ambitious goals. At the same time, there has been a continuing search for larger scale mechanisms to link, support, and strengthen these smaller scale, geographically separate efforts within particular regions or states.

Although sharing the CLT approach to ownership and structure, these local organizations differ from each other in a number of ways including their origins, the sizes of their geographic communities, the specific problems and needs addressed, the resources available to them, and the types of housing programs that have resulted. In this chapter, we illustrate the range of variations by examining three types of situations: (a) isolated communities, where severe conditions of poverty have been addressed by CLTs with a strong emphasis on self-help and voluntarism; (b) the state of Vermont, where an unusual set of state programs has allowed CLTs to create and preserve a variety of affordable housing opportunities; and (c) western resort communities, where CLTs have been developed to create and preserve a supply of affordable owner-occupied housing in the face of extremely high real estate prices. In each case, we also look at factors that provide some degree of linkage and support for the separate local organizations.

## Access to Land and Housing for Low-Income People in Isolated Communities

Two of the earliest rural CLTs, Woodland Community Land Trust in northeast Tennessee and Covenant Community Land Trust in eastern Maine, followed closely in the New Communities tradition. Both have sought to provide affordable housing and tried to reestablish opportunities for very low-income people to pursue land-based activities such as small-scale agriculture and fuel wood production.

Woodland was formed in 1978 in an isolated valley of the Cumberland Mountains in eastern Tennessee, where corporate land ownership and extensive strip mining had resulted in environmental degradation, unemployment in communities where many once had been employed in deep mines, and limited access to land for people who once had been relatively self-sufficient on the land. Woodland (originally named the Clairfield Community Land Association) was established through the efforts of longtime organizer Marie Cirillo, who saw the CLT model not simply as a way in which to provide housing but also as a way in which to address the fundamental economic and environmental problems of Appalachian communities (ICE, 1982, pp.

48-67). As Woodland was being organized, Cirillo also played a role in forming a regional land trust as a vehicle for gaining control of land that could then be made available to locally controlled CLTs.[3] Although the regional land trust has led to the development of only a few CLTs in other communities, Woodland has remained active since inception. It has had a significant impact in the limited setting of Roses Creek Hollow, providing for the construction of new homes, a community garden, and value-added forest products.

Covenant was established by HOME Co-op, a unique organization that began as a marketing vehicle for local craftspeople and has expanded its program to address a broad spectrum of needs experienced by very low-income people in the thinly populated region east of Penobscot Bay (ICE, 1982, pp. 62-75). The efforts of HOME and Covenant are notable for the intense commitment of the people who organized them, their success in attracting large numbers of volunteers from outside the area, and the integration of housing with other programs. Like Woodland, Covenant has built very inexpensive new homes (23 completed by 1995, with 8 more in development that year), using volunteer and "sweat equity" labor and local lumber (HOME operates its own sawmill).

In 1989, HOME and Covenant were instrumental in gaining passage of state legislation providing financing for CLT land acquisition and development through the Maine State Housing Authority. The legislation has encouraged the formation of small, housing-oriented CLTs in other Maine communities. This trend led, in turn, to the 1993 incorporation of a statewide association of CLTs and supporting members, the Maine Homestead Land Trust Alliance. The purpose of the alliance is to promote the CLT approach, acquire and hold land in areas where there are no operating CLTs, and provide support and technical assistance to existing Maine CLTs, most of which are otherwise unstaffed.

## Vermont: A State Mandate
## for Permanent Affordability

Perhaps the most active and successful regional network of CLTs is to be found in Vermont, where CLTs are more nearly the rule than the exception among nonprofit housing organizations. Setting the stage for this trend was the early success of the Burlington Community Land Trust (founded in 1984 with strong support from Burlington's progressive city government) in a state where other nonprofit housing models had not already become strongly established.

In 1987, the CLT movement received an important boost from the Vermont Housing and Conservation Fund, established by the state legislature through the efforts of the remarkable Housing and Conservation Coalition ("Brattleboro Area CLT," 1990, p. 11; "The Vermont Housing and Conservation Fund," 1990, pp. 3-5; "The Walter Smith Farm," 1990, pp. 14-15).

The fund was intended to address not only the environmental impact of development pressures within the state but also the impact of these same pressures on housing affordability. With an emphasis on stewardship and a concern with preserving affordable access for future generations, the newly formed Housing and Conservation Board and progressive staff hired to administer the fund made it clear that housing programs would need to provide for perpetual affordability if they were to receive funding. In a state where local control is highly valued, the board and staff also recognized the importance of supporting the development of local housing organizations. Consequently, several CLTs were established that serve single rural counties or portions of contiguous counties, with capacity-building grants, as well as project funding, supplied by the Housing and Conservation Fund.

The policies of the Housing and Conservation Fund also have influenced policies governing the allocation of other housing subsidies within the state. Thus, the state's federal Community Development Block Grant and HOME program funds have gone only to those housing projects providing for perpetual affordability, which has tended to mean the projects of CLTs. Moreover, the Vermont Housing Finance Agency (VHFA) allows for the mortgaging of resale-restricted, owner-occupied homes with ground leases or deed restrictions and has established a "Perpetually Affordable Housing Program" that provides mortgages at reduced rates only for lower-income purchasers of resale-restricted, perpetually affordable homes.

With the availability of these resources and a significant degree of guidance from the Housing and Conservation Fund, Vermont CLTs have carried out a variety of projects addressing a wide range of housing needs. They have developed homeownership opportunities. They have provided long-term rental housing for very low-income people and people with special needs. They have worked in collaboration with local Habitat for Humanity chapters, specialized service providers, and conservation land trusts. And they have purchased mobile home parks to provide long-term security and affordability for mobile home owners.

In addition, Vermont CLTs have participated in the state's distinctive Homeland Program, through which the Housing and Conservation Fund provides funding to CLTs to subsidize the acquisition of land beneath homes purchased by lower-income households in conventional real estate markets. The homes must be approved by the CLTs, and the prices must be such that, with land costs subsidized, the homes can be financed affordably with VHFA mortgages. The homebuyers receive long-term ground leases from the CLTs with resale restrictions that will preserve affordability perpetually, from one purchaser to the next.

This "buyer-driven" program, which uses the CLT's essential function as long-term land steward and preserver of affordability but which does not involve the CLT in the role of housing developer, has proven popular within the state and has attracted interest from others (e.g., the State of Connecticut has established a similar program). It should be noted, however, that although the program is buyer driven in the case of the initial buyers, when the homes are resold, the CLTs bear responsibility for seeing that they are transferred on the required terms to income-eligible households. The future usefulness and marketability of this pool of homes is a concern that necessarily affects the setting of criteria for determining which homes should be accepted into the program initially. Therefore, it will be important to observe the performance of this scattered site approach over time. At this point, however, it appears to have significant potential as an unobtrusive, cost-effective way in which to accumulate a supply of perpetually affordable housing in rural areas where high land costs are pricing housing beyond the reach of local people.

## New Affordable Housing in
## Western Resort Communities

A small but growing number of communities in the Mountain West and Northwest areas have adopted the CLT model as a way in which to create and preserve a supply of affordable housing in markets where land is expensive and housing is in very short supply. CLTs established in the San Juan Islands of Washington State; Jackson Hole, Wyoming; and Larimer County, Colorado, have attracted interest as possible models for other western communities facing comparable development pressures. Although the circumstances faced by such communities are similar in some ways to those in Vermont, they tend to be highly concentrated in isolated localities and less characteristic of conditions throughout an entire state. Western CLTs have not had the range of resources and guidance provided to Vermont CLTs by the Vermont Housing and Conservation Fund, nor have they undertaken the variety of projects. To date, all of their projects have involved construction of new single-family homes to meet acute shortages of homeownership opportunities for lower-income community residents.

The San Juan Islands provide a dramatic example of a situation where absolutely limited supplies of land are greatly sought by second-home buyers and by people whose economic independence allows them to purchase permanent homes wherever they choose. The resulting high real estate prices have made it extremely difficult for other people to afford decent permanent homes, including longtime residents working in many of the jobs that are essential to the community such as teachers and police.

One of the San Juan Island CLTs is located on Lopez Island, where an acre of land can cost from $75,000 to $125,000 and the average worker's income in 1993 was only $14,000 annually. Lopez Community Land Trust has completed two projects with parcels of a little more than an acre and seven small, single-family units on each parcel. Substantial subsidies were provided by the State of Washington, and construction was completed with the help of large contributions of volunteer and sweat equity labor. Both projects are owned and financed as limited-equity cooperatives, with limitations on the sale of co-op shares established in the co-op documents and reinforced in the ground lease between the co-op and the CLT. Upon sale, the homes will remain affordable for very low-income households.

On nearby Orcas Island, OPAL (*of people and land*) Community Land Trust has taken a somewhat different approach. Financing for OPAL's first construction of 18 single-family homes on a 7-acre parcel was provided through a Rural Housing Service (RHS) demonstration project, with RHS (formerly Farmers Home Administration [FmHA]) loans to the low-income homebuyers financing construction costs and remaining in place as long-term mortgages. (These loans covered a portion of land costs as well because RHS regulations define specific procedures for appraising and mortgaging the value of the leasehold interest held by the lessee-homeowner.) OPAL is now launching a second project on the same model, again to be completed with RHS financing.[4]

In Jackson Hole, the Jackson Hole Community Housing Trust (JHCHT) was established to address an extreme shortage of affordable housing in a market where a limited supply of developable land is highly desired by wealthy purchasers for the construction of luxury homes (the average price of a home was more than $560,000 in 1994; Egan, 1994). The CLT was founded with strong support from local government and local businesses, for whom the supply and cost of labor clearly have been affected by the shortage of affordable housing. In its first project, JHCHT developed 36 single-family units on 4.5 acres of land purchased for $450,000 by Teton County and leased to the CLT, which subleases lots to the homebuyers. As with all CLTs, the ground leases between the CLT and the homeowners require continued owner occupancy and restrict resale prices to preserve affordability. Mortgage financing was provided for some homebuyers by the Wyoming Housing Finance Agency (with Federal Housing Administration insurance) and for the balance of the buyers by RHS ("First Wyoming CLT," 1994, pp. 12-13).

JHCHT has since developed a second project on a donated parcel of land, with a portion of the development cost covered through an unusual financing technique. CLTs normally charge a ground lease fee that includes three components: the cost of taxes on the land, a small administrative charge, and a land use charge. The land use charge is a variable amount that may be minimized or forgiven for low-income homeowners, but some CLTs charge an amount that yields a modest stream of net income over time. JHCHT has "securitized" the stream of income to be realized from ground lease fees and sold the resulting product to a bank.[5] Thus, it is able to capture the future value of these fees at the time of development and use the proceeds to help pay development costs. It should be emphasized that most CLTs retain—and *should* retain—their claim to the monthly lease fee income to help sustain their long-term operations, but in situations where there is an extreme need for production of affordable housing, it can make sense to "pull in" lease fees and use them to help meet the immediate need.

In Larimer County, extreme land and housing costs might not be quite as prevalent or locally concentrated as in the San Juan Islands or Jackson Hole, but the problems affect a greater number of people and communities. A scenic and highly livable environment has attracted huge new populations—both vacationers and permanent residents—that are stretching the existing supply of affordable housing. In response, the Larimer County Community Land Consortium (LCCLC), based in Fort Collins, was established through the efforts of The Resource Assistance Center for Nonprofits (TRAC), which serves as the developer of LCCLC housing.[6] The first project launched by this nonprofit partnership within the city of Fort Collins involved construction of 10 single-family homes that were sold, with ground leases containing typical CLT resale restrictions, to families with incomes between 30% and 60% of the area median income.

In this project and in their plans for future projects, TRAC and LCCLC have placed a strong emphasis on community-building and leadership development activities in the new neighborhoods created by the projects. A separate "chapter" is incorporated for each project, as an affiliate of LCCLC, to serve as a vehicle for leadership training and as a type of homeowners association for the new homebuyers and their neighbors. LCCLC also is exploring the possibility of a statewide organization that would provide training, development assistance, and other support services to small CLTs in scattered rural communities—an organization that would be comparable, in some ways, to the Maine Homestead Land Trust Alliance.

## Conclusion

This brief review of several types of situations where CLTs have been established suggests that these organizations can provide potentially important benefits. The CLT offers an effective way in which to preserve supplies of affordable housing for lower-income households in rural communities experiencing rising real estate prices driven upward by new or seasonal residents whose affluence has little relationship to the local economy. At the same time, by putting significant planning and ownership functions in the hands of local people, the CLT offers a vehicle for leadership development and community revitalization that may help to lead rural communities out of the economic doldrums in which so many now languish.

However, these goals cannot be achieved on a meaningful scale without substantial resources including the financial resources and technical expertise required to develop affordable housing in the first place and the leadership, management skills, and operating support that are needed for long-term success. Such resources are difficult for small organizations in isolated rural communities to mobilize and sustain. The type of statewide support organization that has been established in Maine and is being contemplated in Colorado can help to overcome this difficulty, as can forward-looking, well-integrated state policies such as those in Vermont.

# Notes

1. The Institute for Community Economics (initially called the International Independence Institute) was founded by Ralph Borsodi and Bob Swann in 1967. The origins of the CLT model are described by Swann in an interview published in *Community Economics* ("Bob Swann: An Interview," 1992, pp. 3-5).

2. Information regarding New Communities can be found in *The Community Land Trust: A Guide to a New Model for Land Tenure in America* (International Independence Institute, 1972, pp. 16-25) and *The Community Land Trust Handbook* (ICE, 1982, pp. 39-47).

3. The regional land trust concept implemented in Appalachia was advocated by Bob Swann, ICE's founding director. The basic concept is described in The International Independence Institute's *The Community Land Trust* (1972, p. 35).

4. Information regarding the Lopez and OPAL CLTs can be found in *Community Economics* ("In the San Juan Islands," 1993, pp. 6-7).

5. Telephone interview with Jess Lederman, then executive director of JHCHT, May 24, 1995. Lederman, a retired investment banker, is the architect of this lease fee securitization plan.

6. Interviews with Louise Stitzel and Larry Dunn of the Resource Assistance Center, conducted in Fort Collins, July 20-21, 1995.

# References

Adams, R. (1994, Summer). Director's column. *House Calls,* p. 4.

American Association of Retired Persons. (1994, Summer). Redrawing America. *AARP Bulletin,* pp. 1, 11-12.

American Bankers Association & Independent Bankers Association. (1995, May). [News release]. Washington, DC: Author.

American Business Association and IBAA. *ABA, IBAA announce program to redirect the Farm Credit System to be a wholesale provider of funds for rural banks.* [news release]

Anders, G. (1993, September-October). Preservation of FmHA housing projects. *Housing Law Bulletin,* pp. 79-89. (National Housing Law Project, Oakland, CA)

AuCoin, L. (1979). *Statement by Congressman Les Au Coin on House floor* (reprinted in *Congression Record, 96th Congress, 1st Session,* December 19, 1979, Vol. 125, Part 28). Washington, DC: Government Printing Office.

Baugher, E., & Lamison-White, L. (1996). *Poverty in the United States: 1995* (Current Population Reports, P60-194). Washington, DC: U.S. Department of Commerce, Bureau of the Census.

Bob Swann: An interview. (1992, Summer). *Community Economics,* pp. 3-5. (Springfield, MA: Institute for Community Economics)

Border benefits: Texas towns deserve fair hearing. (1994, April 9). *Dallas Morning News.*

Border Low Income Housing Coalition. (1993). *Border Housing and Community Development Partnership: A proposed partnership of government, community and individuals to improve housing, community and economic conditions in the colonias and barrios of the Texas-Mexico border area.* San Antonio, TX: Author.

Brattleboro Area CLT. (1990, Summer). *Community Economics,* p. 11. (Springfield, MA: Institute for Community Economics)

Carr, J. H., & Megbolugbe, I. F. (1993). The Federal Reserve Bank of Boston study on mortgage lending revisited. *Journal of Housing Research, 4*(2), 277-313.

Chapa, J., & Eaton, D. (directors). (1997). *Colonias housing and infrastructure: Population and housing characteristics, future growth, and housing water and wastewater needs* (3 vols.). Austin, TX: University of Texas, LBJ School of Public Affairs.

Commission on Agricultural Workers. (1992). *Report of the Commission on Agricultural Workers.* Washington, DC: Government Printing Office.

Communications Group. (1993). *Technical assistance to colonias: Final report* (prepared for U.S. Department of Housing and Urban Development). El Paso, TX: Author.

Congressional Budget Office. (1993a). *The Federal Home Loan Banks in the housing finance system.* Washington, DC: Author.

Congressional Budget Office. (1993b). *The housing finance system and federal policy: Recent changes and options for the future.* Washington, DC: Author.

Copley International Corporation. (1979). *Urban and rural housing credit market differentials: Summary volume.* Washington, DC: U.S. Department of Housing and Urban Development.

Day, S. (1978). Housing research relevant to rural development. In *Quality housing environment for rural low-income families* (workshop proceedings). Knoxville, TN: Southern Rural Development Center.

Deavers, K. L. (1991). *Symposium on rural development: Rural America faces many challenges.* Washington, DC: General Accounting Office, Resources, Community, and Economic Division.

Dunn, W. (1987). Regions. *American Demographics, 9*(1), 50-53.

Earhart, C. C., Weber, M. J., & McCray, J. W. (1994). Life-cycle differences in housing perspectives of rural households. *Home Economics Research Journal, 22,* 304-323.

Economic Research Service. (1991). *Rural conditions and trends* (supplement). Washington, DC: U.S. Department of Agriculture.

Economic Research Service. (1995). *Changing financial markets* (USDA/ERS briefing). Washington, DC: U.S. Department of Agriculture.

Egan, T. (1994, November 16). The rich are different: They can afford homes. *The New York Times.*

El Paso comes up short in public aid. (1994, April 10). *El Paso Times.*

Fannie Mae. (1994a). *1994 annual report.* Washington, DC: Author.

Fannie Mae. (1994b). *Showing America a new way home: 1994 report.* Washington, DC: Author.

Farmers Home Administration. (1979). *Summary of FmHA Section 515 rural rental housing caseload as of August 20, 1979* (computer run from MISTR system). Washington, DC: U.S. Department of Agriculture.

Farmers Home Administration. (1981). *A brief history of Farmers Home Administration.* Washington, DC: U.S. Department of Agriculture.

Federal Reserve System. (1995). *Community Reinvestment Act fact sheet.* Washington, DC: Author.

First Wyoming CLT: Utilizing FHA and FmHA financing. (1994, Spring). *Community Economics,* pp. 12-13. (Springfield, MA: Institute for Community Economics)

Freddie Mac. (1994). *Expanding opportunities in your community* (Publication No. 40). Washington, DC: Author.

Gabbard, S., Kissam, E., & Martin, P. L. (1993). The impact of migrant travel patterns on the undercount of Hispanic farmworkers. In *1993 Research Conference on Undercounted Ethnic Populations: Proceedings.* Washington, DC: Government Printing Office.

Governor's Border Working Group. (1993). *Report on Texas colonias.* Austin, TX: Author.

Gruber, K. J., Shelton, B. B., & Hiatt, A. R. (1988). The impact of the presence of manufactured housing on residential property values. *Real Estate Appraiser and Analyst, 54*(2).

Hamilton Securities Group, National Multi-Housing Council, & National Apartment Association. (n.d.). *A report on the multifamily mortgage industry.* Washington, DC: Authors. (Text states that research was conducted in 1993)

Harrison, P. (1994, December). Farmworker housing in crisis: How rural communities can learn from the Arvin experience. *California Agriculture,* pp. 18-22.

Harrison, P. (1995). Safe, clean, and affordable: California farmworker housing needs. *Journal of Architectural and Planning Research, 12*(1), 19-33.

Hartman, H. (1994, October 2). Jobs are there, housing isn't. *Des Moines Register,* pp. G1-G2.

Harvard University Joint Center for Housing Studies. (1997). *The state of the nation's housing 1997.* Cambridge, MA: Harvard University.

Hays, R. A. (1985). *The federal government and urban housing: Ideology and change in public policy.* Albany: State University of New York Press.

Highway boss can't explain disparities. (1994, April 10). *El Paso Times.*

Hombs, M. E. (1990). *American homelessness: A reference handbook.* Santa Barbara, CA: ABC-CLIO.

Housing Assistance Council. (1981). *White paper on rural housing credit.* Washington, DC: Author.

Housing Assistance Council. (1985). *The missing money: FHA mortgage insurance in rural areas.* Washington, DC: Author.

Housing Assistance Council. (1986a). *The missing money, Part II: Use of Federal Housing Administration mortgage insurance by rural banks and savings and loan associations.* Washington, DC: Author.

Housing Assistance Council. (1986b). *Who will house farmworkers? An examination of state programs.* Washington, DC: Author.

Housing Assistance Council. (1987). *Displacement of tenants through prepayment of FmHA Section 515 loans.* Washington, DC: Author.

Housing Assistance Council. (1991). *Rural homelessness: A review of the literature.* Washington, DC: Author.

Housing Assistance Council. (1992). *Who will house farmworkers? An update on state and federal programs.* Washington, DC: Author.

Housing Assistance Council. (1994a). *State data sheets: An overview of poverty and housing data from the 1990 census.* Washington, DC: Author

Housing Assistance Council. (1994b). *Taking stock of rural poverty and housing for the 1990s.* Washington, DC: Author.

Housing Assistance Council. (1995). *McKinney Act programs in nonmetro areas: How far do they reach? Part 1: Distribution of funds, 1987-1991.* Washington, DC: Author.

Housing Assistance Council. (1996). *Fitting the pieces together: An examination of farmworker housing data sources.* Washington, DC: Author.

Housing Assistance Council. (1997a). *A brief and selective historical outline of mutual self-help housing in the United States.* Washington, DC: Author.

Housing Assistance Council. (1997b). *The RHS housing program in fiscal year 1996: A year of serious consequences.* Washington, DC: Author.

Housing Assistance Council. (1997c). *Survey of demand for the RHS farm labor housing program.* Washington, DC: Author.

Hunt, M. E., & Ross, L. E. (1990). Naturally occurring retirement communities: A multiattribute examination of desirability factors. *The Gerontologist, 30,* 667-674.

In the San Juan Islands: CLTs combat high land costs. (1993, Fall). *Community Economics,* pp. 6-7. (Springfield, MA: Institute for Community Economics)

Institute for Community Economics. (1982). *The community land trust handbook.* Emmaus, PA: Rodale.

Interagency Council on the Homeless. (1994). *Priority: Home! The federal plan to break the cycle of homelessness.* Washington, DC: U.S. Department of Housing and Urban Development.

Inter-America Research Associates. (1980). *National Farmworker Housing Study: Final report.* Rosslyn, VA: Author.

International Independence Institute. (1972). *The community land trust: A guide to a new model for land tenure in America.* Cambridge, MA: Center for Community Economic Development.

Kennedy, M., Goldberg, D., & Fishbein, A. (1990). *Location, location, location: Report on residential mortgage credit availability in rural areas.* Washington, DC: Center for Community Change.

Kravitz, L., & Collings, A. (1986). Rural housing policy in America: Problems and solutions. In R. Bratt, C. Hartman, & A. Meyerson (Eds.), *Critical perspectives on housing.* Philadelphia: Temple University Press.

Larson, A., & Plascencia, L. (1993). *Migrant Enumeration Project.* Washington, DC: Migrant Legal Action Program.

Lazere, E. B., Leonard, P. A., & Kravitz, L. L. (1989). *The other housing crisis: Sheltering the poor in rural America.* Washington, DC: Center on Budget and Policy Priorities and Housing Assistance Council.

Lower Mississippi Delta Development Commission. (1990). *Final report: The Delta initiatives— Realizing the dream fulfilling the potential* (D. W. Brown, Ed.). Memphis, TN: Author. (Records archived at University of Memphis library)

Lowi, T. J. (1979). *The end of liberalism: The second republic of the United States.* New York: Norton.

Loza, M. (1994). *Testimony before the Subcommittee on Environment, Credit, and Community Development, House Committee on Agriculture* (July 21, 1994). Washington, DC: Housing Assistance Council.

Marans, R. W., & Wellman, J. D. (1978). *The quality of nonmetropolitan living: Evaluation, behaviors and expectations of northern Michigan residents.* Ann Arbor: University of Michigan Press.

Marcuse, P. (1986). Housing policy and the myth of the benevolent state. In R. Bratt, C. Hartman, & A. Meyerson (Eds.), *Critical perspectives on housing.* Philadelphia: Temple University Press.

Margolis, R. J. (1967). *Something to build on: The future of self-help housing in the struggle against poverty.* Washington, DC: International Self-Help Housing Associates and the American Friends Service Committee.

McCray, J. W. (1989). Housing affordability: Concept and reality. *Family Economic Review, 2*(1), 14-18.

McCray, J. W. (1992). Overall models of housing affordability, quality, and diversity. In J. W. McCray & G. E. Shelton (Eds.), *Affordable housing in the rural South: A causal model of barriers and incentives* (Southern Cooperative Series Bulletin No. 371). Pine Bluff: University of Arkansas Agriculture Experiment Station.

McCray, J. W., Cotledge, B. J., Conley, R. D., & Watson, B. J. (1990). *Housing problems and solutions in the Lower Mississippi Delta.* Report to the Lower Mississippi Delta Development Commission, Memphis, TN. (Records archived at University of Memphis library)

McLaughlin, R. W. (1993). *Establishing a primary mortgage market in Indian country.* Washington, DC: National Indian Policy Center.

Mikesell, J. J. (1985). Rural banks reflect the local economy. *Rural Development Perspectives, 2*(1), 17.

Milkove, D. (1990). *The changing nature of the rural finance system: Financial market interventions as a rural development strategy.* Washington, DC: U.S. Department of Agriculture, Economic Research Service.

Milkove, D. L., & Sullivan, P. J. (1990). *Deregulation and the structure of rural financial markets.* Washington, DC: U.S. Department of Agriculture, Economic Research Service.

Morris, E. W., & Winter, M. (1982). Housing. In D. A. Dillman & D. Hobbs (Eds.), *Rural society in the U.S.: Issues for the 1980s.* Boulder, CO: Westview.

Munnell, A. H., Browne, L. E., McEneaney, J., & Tootell, G. (1992). *Mortgage lending in Boston: Interpreting HMDA data* (working paper). Boston: Federal Reserve Bank of Boston.

Murrow, E. R. (producer). (1960, November 25). *Harvest of shame* (film documentary). New York: Columbia Broadcasting System.

National Association of Home Builders. (1992). Mortgage credit in rural America. *Pacific Mountain Review, 10*(2), 11-13. (Rural Community Assistance Corporation)

National Association of Realtors. (1990). *The evolution of mortgage finance in the eighties* (staff study). Washington, DC: Author.

National Commission on American Indian, Alaska Native, and Native Hawaiian Housing. (1993). *Supplemental report and Native American housing improvements legislative initiative.* Washington, DC: Author.

National Rural Housing Coalition. (1995). *The FHFB Affordable Housing Program: Uses in rural areas* (briefing). Washington, DC: Author.

National Task Force on Financing Affordable Housing. (1992). *From the neighborhoods to the capital markets.* Washington, DC: Allstate Insurance Company.

National Task Force on Rural Housing Preservation. (1992). *Preserving rural housing: Final report of the National Task Force on Rural Housing Preservation.* Washington, DC: Housing Assistance Council, California Coalition for Rural Housing Project, and National Housing Law Project.

Office of the Comptroller of the Currency. (1993). *1992 national bank community development survey report.* Washington, DC: Author.

Office of Policy Development and Research. (1995). *Credit gaps in the multifamily mortgage markets.* Washington, DC: U.S. Department of Housing and Urban Development.

Office of the State Auditor of Texas. (1993). *Providing water and wastewater service to the colonias: A shared responsibility.* Austin, TX: Author.

Payne, K. (1994, March). *Documentation log and attitudinal survey of homeless in Grand Island, NE.* Kearney: University of Nebraska.

Pheabus, J. F., & Rowland, R. (Eds.). (n.d.) *Mortgage banking basics.* Washington, DC: Mortgage Bankers Association of America.

Rochin, R. I., & Castillo, M. D. (1993). *Immigration, colonias formation and Latino poor in rural California: Evolving "immiseration."* Los Angeles, CA: Tomás Rivera Center.

Rossi, C. (1991). Rural banks in good shape, S&L's a potential trouble spot. *Rural Development Perspectives, 7*(3), 36-37.

Runsten, D., & Kearney, M. (1994). *A survey of Oaxacan village networks in California agriculture.* Davis: California Institute for Rural Studies.

Savitch, H. V. (1979). *Urban policy and the exterior city: Federal, state and corporate impacts upon major cities.* New York: Pergamon.

Shelton, G. G., & Sweaney, A. L. (1983). *Perceptions of alternative housing: Housing for low- and moderate-income families* (Southern Cooperative Series Bulletin 298). Athens: University of Georgia.

Sindt, R. P., & Guy, D. C. (1985). Economics of rural housing: Challenges and changes. *Housing and Society, 12,* 147-160.

The state of low-income housing. (1997). *Journal of Housing, 54*(6), 16-21.

The structure of the banking and thrift industry in rural areas. (1991). *Rural Conditions and Trends,* Supplement 1, p. 6.

Sullivan, P. J. (1990). *Overview: Financial market intervention as a rural development strategy.* Washington, DC: U.S. Department of Agriculture, Economic Research Service.

Texas Department of Housing and Community Affairs. (1993). *Texas colonias: Creating real solutions to poverty.* Austin, TX: Author.

Unfair funding spurs inquiries. (1994, May 8). *Houston Chronicle.*

Urban Institute. (1990). *Report to Congress on the availability and use of mortgage credit in rural areas.* Washington, DC: Author.

Urban Systems Research & Engineering. (1977). *The barriers to equal opportunity in rural housing markets,* Vol. 1: *Analysis and findings.* Washington, DC: U.S. Department of Housing and Urban Development.

U.S. Bureau of the Census. (1990). *U.S. census of population and housing, 1990.* Washington, DC: Government Printing Office.

U.S. Bureau of the Census. (1992a). Census of population and housing, 1990. Summary tape file 3 on CD-ROM Technical Documentation. Washington, DC: Author.

U.S. Bureau of the Census. (1992b). *Summary population and housing characteristics, United States* (1990 Census of Population and Housing summary). Washington, DC: Government Printing Office.

U.S. Bureau of the Census. (1994). *Residential finance: 1990 census of housing.* Washington, DC: Government Printing Office.

U.S. Bureau of the Census. (1996). *Poverty in the United States: 1995* (Current Population Reports, Series P-60-194). Washington, DC: Government Printing Office.

U.S. Bureau of the Census & U.S. Department of Housing and Urban Development. (1995). *American Housing Survey for the United States: 1993* (Current Housing Reports, H150/93). Washington, DC: Government Printing Office.

U.S. Bureau of the Census & U.S. Department of Housing and Urban Development. (1997). *American Housing Survey for the United States: 1995* (Current Housing Reports, H150/95RV). Washington, DC: Government Printing Office.

U.S. Department of Labor. (1993). *U.S. farmworkers in the post-IRCA period (based on data from the National Agricultural Workers Survey (NAWS).* Washington, DC: Author.

U.S. Department of Labor. (1994). *Migrant farmworkers: Pursuing security in an unstable labor market* (Research Report No. 5, Office of Program Economics). Washington, DC: Author.

U.S. General Accounting Office. (1988). *Rural rental housing: Impact of Section 515 loan prepayments on tenants and housing availability* (briefing report to the chairman, Subcommittee on Housing and Community Development, House of Representatives, GAO/RCED-88-15BR). Washington, DC: Author.

U.S. General Accounting Office. (1991). *Federal Agricultural Mortgage Corporation: Potential role in the delivery of credit for rural housing* (GAO/RCED-91-180). Washington, DC: Author.

U.S. General Accounting Office. (1993). *Rural development: Profile of rural areas* (GAO/RCED-93-40FS). Washington, DC: Government Printing Office.

The Vermont Housing and Conservation Fund. (1990, Summer). *Community Economics,* pp. 3-5. (Springfield, MA: Institute for Community Economics)

Villarejo, D., & Runsten, D. (1993). *California's agricultural dilemma.* Davis: California Institute for Rural Studies.

The Walter Smith farm. (1990, Summer). *Community Economics,* pp. 14-15. (Springfield, MA: Institute for Community Economics)

Wiener, R. (1993). *Crisis in U.S. housing policy: The prepayment problem in FmHA's rural rental housing program.* Unpublished dissertation, Graduate School of Architecture and Urban Planning, University of California, Los Angeles.

# Suggested Readings

Apgar, W. C. (1990). Which housing policy is best? *Housing Policy Debate, 1,* 1-32.

Carey, P. (1995, October). *Mutual self-help housing.* Paper prepared for Fannie Mae Research Roundtable, Washington, DC.

Collings, A. (1995, October). *The role of the federal rural housing programs.* Paper prepared for Fannie Mae Research Roundtable, Washington, DC.

Congressional Budget Office. (1982). *Rural housing programs: Long-term costs and their treatment in the federal budget.* Washington, DC: Author.

Cook, C., & Krofta, J. (1995, October). *Affordable housing in the rural Midwest.* Paper prepared for Fannie Mae Research Roundtable, Washington, DC.

Dolbeare, C. N. (1995, October). *Conditions and trends in rural housing.* Paper prepared for Fannie Mae Research Roundtable, Washington, DC.

Drier, P., & Hulchanski, J. D. (1993). The role of nonprofit housing in Canada and the United States: Some comparisons. *Housing Policy Debate, 4,* 43-80.

Galston, W. A., & Baehler, H. J. (1995). *Rural development in the United States: Connecting theory, practice, and possibilities.* Washington, DC: Island.

Griffith, D., Kissam, E., with others. (1995). *Working poor: Farmworkers in the United States.* Philadelphia: Temple University Press.

Grogan, P. (1995). Letter from LISC's president. *Rural Developments, 1*(1), 2.

Harrison, P. (1995, October). *New housing forms for low-income rural families.* Paper prepared for Fannie Mae Research Roundtable, Washington, DC.

Housing Assistance Council. (1984). *Taking stock: Rural people and poverty, 1970-1983.* Washington, DC: Author.

Housing Assistance Council. (1988). *"A home of our own": The costs and benefits of the rural homeownership programs.* Washington, DC: Author.

Housing Assistance Council. (1993). *Community land trusts and rural housing.* Washington, DC: Author.

Housing Assistance Council. (1994a). *Filling the gaps: A study of capacity needs in rural nonprofit housing development.* Washington, DC: Author.

Housing Assistance Council. (1994b). *Taking stock of rural poverty and housing for the 1990s.* Washington, DC: Author.

Housing Assistance Council. (1995a). *The FmHA housing program in fiscal year 1994: A difficult year.* Washington, DC: Author.

Housing Assistance Council. (1995b). *McKinney Act programs in nonmetro areas: How far do they reach?* Washington, DC: Author.

Housing Assistance Council. (1995c). *The poorest of the poor: Female-headed households in nonmetro America.* Washington, DC: Author.

Housing Assistance Council. (1997). *Rural housing and welfare reform: HAC's 1997 report on the state of the nation's housing.* Washington, DC: Author.

Jencks, C. (1994). *The homeless.* Cambridge, MA: Harvard University Press.

Joint Center for Housing Studies. (1994). *The state of the nation's housing 1995.* Cambridge, MA: Harvard University Joint Center for Housing Studies.

Kennedy, M., Goldberg, D., & Fishbein, A. (1990). *Location, location, location: Report on residential mortgage credit availability in rural areas.* Washington, DC: Center for Community Change.

Lazere, E., Leonard, P., & Kravitz, L. (1989). *The other housing crisis: Sheltering the poor in rural America.* Washington, DC: Center on Budget and Policy Priorities and Housing Assistance Council.

Martin, P. L., & Martin, D. A. (1994). *The endless quest: Helping America's farm workers.* Boulder, CO: Westview.

Martinez, Z. Q., & Kamasaki, C. (1995, October). *Colonias: A policy framework.* Paper prepared for Fannie Mae Research Roundtable, Washington, DC.

McCray, J. C. (1995, October). *Affordable rural housing in the South.* Paper prepared for Fannie Mae Research Roundtable, Washington, DC.

Mikesell, J., & Davidson, S. (1982). Financing rural America: A public policy and research perspective. In *Rural financial markets: Research issues for the 1980s—Proceedings, December 9 and 10, 1982.* Chicago: Federal Reserve Bank of Chicago.

Nelson, K. P. (1994). Whose shortage of affordable housing? *Housing Policy Debate, 5,* 401-442.

Peck, S. (1995, October). *Many harvests of shame: Housing for farmworkers.* Paper prepared for Fannie Mae Research Roundtable, Washington, DC.

Quercia, R., McCarthy, G., & Stegman, M. (1995). Mortgage default among rural, low-income borrowers. *Journal of Housing Research, 6,* 349-369.

Rural Sociological Society. (1993). *Persistent poverty in rural America.* Boulder, CO: Westview.

Stover, M. (1995, October). *Out of sight: The rural homeless.* Paper prepared for Fannie Mae Research Roundtable, Washington, DC.

Strauss, L. (1995, October). *Credit and capital needs for affordable rural housing.* Paper prepared for Fannie Mae Research Roundtable, Washington, DC.

U.S. Department of Agriculture, Economic Research Service. (1995). *Changing rural financial markets: RECD briefing.* Unpublished manuscript.

U.S. Department of Housing and Urban Development. (1995). *The national homeownership strategy: Partners in the American dream.* Washington, DC: Government Printing Office.

Walker, C. (1993). Nonprofit housing development: Status, trends, and prospects. *Housing Policy Debate, 4,* 369-414.

White, K., Lemke, J., & Lehman, L. (1995, October). *Community land trusts and rural housing.* Paper prepared for Fannie Mae Research Roundtable, Washington, DC.

Wiener, R. J. (1995, October). *Preserving rural rental housing: The prepayment issue in FmHA's multifamily program.* Paper prepared for Fannie Mae Research Roundtable, Washington, DC.

Wilson, H. L., & Carr, J. H. (1995, October). *Impact of federal interventions on private mortgage credit in rural areas.* Paper prepared for Fannie Mae Research Roundtable, Washington, DC.

# Index

# About the Authors

**JOSEPH N. BELDEN** is Deputy Executive Director of the Housing Assistance Council (HAC). He was previously Research Associate at HAC and has also worked for the U.S. Department of Agriculture and for several policy-oriented research organizations in Washington, D.C.

**PETER N. CAREY** is Executive Director of Self-Help Enterprises, a non-profit housing and community development organization that serves the San Joaquin Valley in California. He has also served as a planning commissioner, city council member, and mayor of Visalia, California.

**JAMES H. CARR** is Senior Vice President of the Fannie Mae Foundation. He previously served as Assistant Director of Tax Policy with the U.S. Senate Budget Committee and as a Research Associate with the Center for Urban Policy Research at Rutgers University.

**GORDON CAVANAUGH** is a partner in the Washington law firm of Reno and Cavanaugh and General Counsel of the Council of Large Public Housing Authorities. He has also served as Administrator of the Farmers Home Administration; Executive Director of the Housing Assistance Council; and Chairman of the Philadelphia Housing Authority.

**ART COLLINGS** served the USDA Farmers Home Administration as County Supervisor, State Office Official, Agency Training Director, and

Special Assistant to the Administrator. He also has worked as Senior Housing Specialist at the Housing Assistance Council for more than 20 years.

**CHRISTINE C. COOK** is Associate Professor in the housing program, Department of Human Development and Family Studies, at Iowa State University. Her teaching and research interests focus on housing and neighborhood needs and choices of special populations—children with disabilities, elderly, women, and minorities—from a human development perspective.

**SUE R. CRULL** is Assistant Professor in the housing program, Department of Human Development and Family Studies, at Iowa State University. She also has been Associate Director of the Kercher Center for Social Research at Western Michigan University. Her research interests include households at risk and the role of housing in rural development.

**SURABHI DABIR** is Research Associate and Project Manager for the Housing Assistance Council's strategic technical assistance initiatives in the Lower Mississippi Delta and the Southwest border colonias. She also has worked on fair housing issues for Aspen Systems Corporation.

**CUSHING N. DOLBEARE** is a consultant on housing and public policy and has worked on housing issues since 1952. She is founder of the National Low Income Housing Coalition and its sister organization, the Low Income Housing Information Service. She has also served as Executive Director of the National Rural Housing Coalition and as interim Executive Director of the National Coalition for the Homeless.

**PATRICIA HARRISON** is Associate Professor of Environmental Design at the University of California at Davis. She conducts research in low-income housing and works with nonprofit housing groups on development of housing projects. Two of her low-income architectural projects are Village Park and Crossroads Gardens, both in Sacramento, California.

**CHARLES KAMASAKI** has served as Vice President for Research, Advocacy and Legislation for the National Council of La Raza (NCLR) since 1989. He previously was Director of NCLR's Policy Analysis Center, a position he held for 7 years.

**JANET A. KROFTA** is Senior Research Analyst in Housing Services at Aspen Systems Corporation. Her most recent work focuses on consolidated plan summaries, nonprofit organizations, and affordable housing mobility. She also has conducted research for the Housing Assistance Council and taught housing at the University of Minnesota.

**MICHAEL LEHMAN** is a consultant to nonprofit groups on community development. Previously, he was a technical assistance provider for the Institute for Community Economics. For more than a decade, he has worked with community-based efforts to develop affordable housing.

**JILL LEMKE** is a planner in Buffalo, New York. While coauthoring the chapter on community land trusts in this book, she was a graduate student in the Department of City and Regional Planning at Cornell University and a researcher for the Institute for Community Economics.

**ZIXTA Q. MARTÍNEZ** is Program Director of the National Fair Housing Alliance. Under her leadership, new fair housing agencies have been organized in several states. She also has conducted research for the National Council of La Raza.

**JACQUELYN W. McCRAY** is Dean/Director for 1890 Research and Extension in the School of Agriculture, Fisheries, and Human Sciences at the University of Arkansas at Pine Bluff. She is a past president of the American Association of Housing Educators and author of more than 40 refereed publications in major housing and rural development research outlets.

**SUSAN PECK** is Director of the Housing Assistance Council's Western Regional Office in California. In 1972, she began her career with HAC as a Research Associate. She administers HAC's "Increasing Access to Rural Housing Service's Housing Programs" projects.

**MARY STOVER** is Community Housing Development Organization (CHDO) Project and Conference Director with the Housing Assistance Council. She joined HAC in 1992 as a Research Associate and previously served as a Program Associate for the Shaping Growth in American Communities program of Partners for Livable Places.

**LESLIE R. STRAUSS** is Director of Research and Information at the Housing Assistance Council. Before joining HAC, she practiced real estate law with Arnold & Porter. She held research-related positions with Rural America, the National Rural Housing Coalition, and the Tug Valley Recovery Center in Williamson, West Virginia.

**KIRBY WHITE** is Technical Assistance Director for the Institute for Community Economics (ICE) and was also editor of ICE's periodical *Community Economics*. He has been associated with ICE and the community land trust movement since 1981.

**ROBERT J. WIENER** is Executive Director of the California Coalition for Rural Housing Project, a statewide coalition of builders and advocates who support housing production and preservation for low-income and rural Californians. He is also a lecturer at the University of California at Davis, where he teaches housing policy and community research methods.

**HAROLD O. WILSON** is a Senior Program Director for the national Rural Local Initiatives Support Corporation. Previously, he was Executive Director of the Housing Assistance Council, Executive Director of Rural Housing Improvement, Inc., and Vice President of the Cooperative Housing Foundation.